Spatial
Cloud
Computing

A Practical Approach

Spatial
Cloud
Computing

A Practical Approach

Chaowei Yang
Qunying Huang

With the collaboration of

Zhenlong Li
Chen Xu
Kai Liu

CRC Press
Taylor & Francis Group
Boca Raton London New York

CRC Press is an imprint of the
Taylor & Francis Group, an **informa** business

CRC Press
Taylor & Francis Group
6000 Broken Sound Parkway NW, Suite 300
Boca Raton, FL 33487-2742

First issued in paperback 2017

ISBN 13: 978-1-138-07555-9 (pbk)
ISBN 13: 978-1-4665-9316-9 (hbk)

Library of Congress Cataloging-in-Publication Data

Yang, Chaowei.
  Spatial cloud computing : a practical approach / authors, Chaowei Yang, Qunying
    pages cm
  Summary: "A computing cloud is a set of network enabled services, providing scal
inexpensive computing platforms on demand, which can be accessed in a simple way
book helps its readers understand the process of how to deploy and customize geospa
applications onto clouds, as well as how to optimize clouds to make them better sup
geospatial applications. It also discusses and presents the strategies for customizing
types of applications to better utilize the cloud capabilities, such as on-demand servi
Provided by publisher.
  Includes bibliographical references and index.
  ISBN 978-1-4665-9316-9 (hardback)
  1. Geospatial data. 2. Cloud computing. I. Huang, Qunying. II. Title.

G70.217.G46Y36 2014
004.67'82--dc23   2

Visit the Taylor & Francis Web site at
http://www.taylorandfrancis.com

and the CRC Press Web site at
http://www.crcpress.com

Contents

Preface

WHY DID WE WRITE THIS BOOK?

There are several motivations that led to the writing of this book. We started utilizing cloud computing for geoscience applications around 2008, when cloud computing was just starting to take shape. During the past several years, many cloud computing related books have been published in the computer science domain. But there is no such book detailing the various aspects of how the geoscience community can leverage cloud computing. Our first motivation with this book was to fill this gap to benefit the geoscience community to cover the various aspects of why and how to adopt cloud computing for geosciences (Parts I and II).

Our second motivation came from our well-cited 2011 spatial cloud computing publication of the *International Journal of Digital Earth*. The paper introduced the general concepts and the benefits that can be brought about by cloud computing to geoscience research and applications. We also received inquiries on how to achieve those benefits and how to use cloud computing in a pedagogical fashion. This book in one aspect responds to the requests with Parts II and III on how to cloud-enable geoscience applications step by step.

We have conducted a series of research and development initiatives for using cloud computing for geoscience applications. The projects, detailed in Parts II, III, and IV, range from migrating a Web portal onto a cloud service to investigating the readiness of cloud computing for geosciences using both commercial cloud services and open-source solutions. We also believed that firsthand experience would be very useful if documented in a systematic fashion for geoscientists and geoscience application developers to evaluate, select, plan, and implement cloud operations for their geoscience applications. This was our third motivation that enlightened us to write this book.

We combined our experience gained during the past six years to write this systematically progressive book for demonstrating how geoscience communities can adopt cloud computing from concepts (Part I), migrating applications to cloud services (Part II), cloud-enabling geoscience applications (Part III), cloud readiness tests and federal cloud-adoption approaches (Part IV), and the future research direction of cloud computing for geosciences (Part V). We expect this book to provide systematic knowledge for readers who wish to get a sense of spatial cloud computing, adopt cloud computing for their applications, or conduct further research in spatial cloud computing.

HOW DID WE WRITE THE BOOK?

In 2012, CRC Press/Taylor & Francis (Irma Britton) saw the need for a cloud computing book for the geoscience communities and agreed with the author team to materialize such an effort. During the past year, we followed 13 steps to ensure a well-written and structured book for our audience: (1) Drs. Chaowei Yang, Qunying Huang, Chen Xu, and Mr. Zhenlong Li and Mr. Kai Liu worked to define the structure and content of the book with each of the chapter authors, who are the developers and researchers for relevant projects. (2) Each chapter was initially written by the authors with the participation of one or several editors. (3) To ensure that the content of each chapter corresponded to the overall book design, Yang was responsible for the review of each chapter in Parts I, II, and V; Xu was responsible for the review of Part IV; and Li was responsible for the review of Chapters 4, 5, and Part III. (4) Structural and content comments were provided to the authors of each chapter to ensure that the overall organization of the book was integrated. (5) Authors of each chapter revised and reviewed the entire chapter by themselves. (6) An internal review of a chapter by authors of other relevant chapters was conducted to ensure the smooth flow of chapters. (7) Authors of each chapter revised and restructured the book chapter as needed. (8) Each chapter was sent out for review by two to four external reviewers. (9) Authors of each chapter and section (part) editors collaborated to address the external review comments. (10) Yang, Xu, and Li did a final review and proof of the chapters. (11) The chapters and the entire book were finalized with Taylor & Francis editors after being formatted by Nanyin Zhou. (12) Huang, Yang, Li, Xu, and Liu worked together to develop the online content including lecture slides for each chapter and online code, scripts, virtual machine images, videos, and documents for readers to easily repeat the cloud deployment and migration processes described in the book. (13) The online content is published on the Taylor & Francis Web site for the book. The book is written by authors who have firsthand experience to ensure the content is well covered. The content was also checked to ensure its organization as a single volume with the project's team leaders (all editors) and principal investigator (Yang) reviewing and approving all content.

WHAT IS THIS BOOK ABOUT?

This book comprehensively introduces knowledge of spatial cloud computing through practical examples in 17 chapters from 5 aspects including: (a) What are the essential cloud computing concepts and why do geosciences need cloud computing? (b) How can simple geoscience applications be migrated to cloud computing? (c) How can complex geoscience applications be cloud-enabled? (d) How can a cloud service be tested to see if it is ready to support geoscience applications? (e) What are the research issues in need of further investigation?

Part I introduces the geoscience requirements for cloud computing in Chapter 1, summarizes the architecture, characteristics, and concepts of cloud computing in Chapter 2, and discusses the enabling technologies of cloud computing in Chapter 3.

Part II introduces the general procedures and considerations when migrating geoscience applications onto cloud services. Chapter 4 demonstrates how to use cloud services through deploying a simple Web application onto two popular cloud services: Amazon EC2 and Windows Azure. Chapter 5 introduces the common procedures for deploying general geoscience applications onto cloud platforms with needs for server-side scripting, database configuration, and high performance computing. Chapter 6 discusses how to choose cloud services based on general cloud computing measurement criteria and cloud computing cost models.

Part III demonstrates how to deploy different geoscience applications onto cloud services. Chapter 7 explains how users can interact with cloud services using ArcGIS in the Cloud as an example. The other three chapters demonstrate how consumers can cloud-enable three different complex geoscience applications: (1) cloud-enabling databases, spatial index, and spatial Web portal technologies to support GEOSS Clearinghouse, (2) cloud-enabling stand-alone model simulations and model output visualization for Climate@Home, and (3) leveraging elastic cloud resources to support disruptive events (e.g., dust storm) forecasting.

Part IV examines the readiness of cloud computing to support geoscience applications using open-source cloud software solutions and commercial cloud services. Chapter 11 introduces and compares three commercial cloud services: Amazon EC2, Windows Azure, and Nebula. In Chapter 12, the readiness of these three cloud services are tested with the three applications described in Part III. Chapter 13 introduces four major cloud computing open-source solutions including CloudStack, Eucalyptus, Nimbus, and OpenNebula; their performance and readiness are tested and compared in Chapter 14. Chapter 15 presents the background, architecture design, approach, and coordination of GeoCloud, which is a cross-agency initiative to define common operating system and software suites for geoscience applications.

Finally, Part V reviews the future research and developments for cloud computing in Chapters 16 and 17. Chapter 16 introduces data, computation, concurrency, and spatiotemporal intensities of geosciences and how cloud services can be leveraged to solve the challenges. Chapter 17 introduces the research directions from the aspects of technology, vision, and social dimensions.

ONLINE CONTENT OF THE BOOK

To help readers better use this book for different purposes, the following online content is provided at: http://www.crcpress.com/product/isbn/9781466593169.

- *Lecture slides for each chapter*—To serve educational purposes, this book provides slides for instructors to assist them in teaching the content. The slides are closely mapped to the book chapter content.
- *Key questions*—Five to ten questions that lead a reading of the book are available for each book chapter. The answers for those questions can be found through the context of the chapters as a review of the core content.
- *Virtual machine images of the application examples used in this book*—Chapters 4, 5, 7, 8, 9, and 10 include different levels of examples, from a simple Web application to complex geoscience applications, such as GEOSS Clearinghouse (Chapter 8), Climate@Home (Chapter 9), and dust storm forecasting (Chapter 10). The images contain the source code and data for those examples available for Amazon EC2. Therefore, audiences can directly launch cloud virtual machines from those images and test those examples.
- *Manuals for deploying the application examples*—Details of deploying workflow applications onto cloud services are included (Chapters 4, 5, 7, 8, 9, and 10). In addition, Chapters 12 and 14 also include the detailed workflow for testing the cloud services.
- *Scripts for installing and configuring application examples and cloud services.*
- *Videos to show step-by-step deployment of the application examples.*

WHO IS THE AUDIENCE?

To thoroughly understand spatial cloud computing, especially in supporting the computing needs of geoscience applications, we wrote this book based on our last decade's investigation into many projects in collaboration with a variety of agencies and companies to solve the computing problems of geoscience applications. The reading of the book should progress

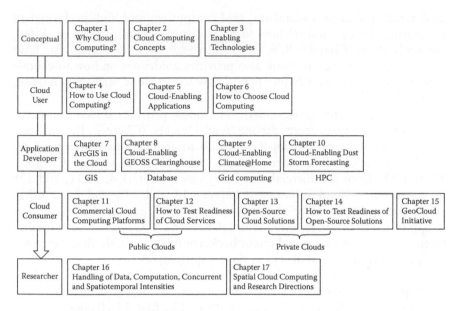

Figure P.1 Reading guide of the book.

in the sequence of the parts and book chapters. But some of them can be omitted based on interest. Figure P.1 depicts the workflow of the chapters for a reader in the knowledge progression sequence.

This book can be used as follows:

(1) As a textbook by professors and students who plan to learn different aspects of cloud computing with the combination of the online slides and examples for class lectures. Each chapter includes lecture slides and is appropriate to serve as independent lecture content. The chapters of Parts II to IV include detailed examples, source code, and data, which could be used for class practice to provide students with hands-on experiences of cloud usage and deployment. These examples can also be used as homework to reinforce what students learned from the lecture. In addition, the examples are carefully selected and considered ranging from simple to complex so that students with different levels of background can follow along. Five to ten questions are provided for each chapter to help students summarize the core content of the respective chapter.

(2) A manual for cloud-enabled application developers with the guidelines is progressively provided in Parts II, III, and IV. This book first provides a general guideline of how to deploy applications onto cloud services (Chapter 4). And then based on the guideline, a common workflow for deploying geoscience applications onto cloud services is introduced (Chapter 5). Based on this common workflow, three practical examples are used to demonstrate (a) how to cloud-enable three different types of

geoscience applications (database, grid computing, and high performance computing [HPC]), and (b) how to handle special requirements of different applications (Chapters 8, 9, and 10). In addition to demonstrating how to use cloud services, this book also provides guidelines on how to choose suitable cloud services (Chapter 6) and how to test cloud services (Chapters 12 and 14).

(3) A reference for geoscientists. The book provides different aspects of cloud computing, from driving requirements (Chapter 1), concepts (Chapter 2), and technologies (Chapter 3) to applications (Chapters 8, 9, and 10), from cloud provider selection (Chapter 6) to testing (Chapters 12 and 14), from commercial cloud services (Chapters 4, 5, 11, and 12) to open-source cloud solutions (Chapters 13 and 14), and from using cloud computing to solve contemporary research and application issues (Chapter 16) to future research topics (Chapter 17). Geoscientists with different research and science domain backgrounds can easily find the cloud computing knowledge that will fit their requirements.

(4) A reference for general IT professionals and decision makers. This book provides references to the concepts, the technical details, and the operational guidelines of cloud computing. The first 15 chapters provide incremental descriptions about different aspects of cloud computing. Chapters 4, 5, and 7 through 15 are closely related to daily IT operations. Decision makers can use Chapters 1 to 3 to build a foundational understanding of cloud computing; then skip to Chapter 6 for considerations related to cloud service selection; and find useful information in Chapters 11, 13, and 15, which cover both commercial and private cloud introductions and evaluations that are most relevant to their decision making.

Acknowledgments

The authors have gained their experience from participating in projects totaling over $10 million in funding from a variety of agencies and companies including the National Aeronautics and Space Administration (NASA) OCIO, NASA MAP Program, NASA Applied Sciences Program, NASA High End Computing Program, National Science Foundation (NSF) EarthCube Program, NSF I/UCRC Program, NSF CNF Program, and NSF Polar Programs, U.S. Geological Survey (USGS) SilvaCarbon Program, Federation of Earth Science Information Partners (FGDC), Microsoft Research, intergovernmental Group on Earth Observation (GEO), ESIP, Association of American Geographers (AAG), Cyberinfrastructure Specialty Group (CISG), ISPRS WG II/IV, the international association of Chinese Professionals in Geographic Information Sciences (CPGIS), and Amazon Inc. The nonauthor project collaborators include Mirko Albani, John Antoun, Jeanette Archetto, Jeff de La Beaujardiere, Don Badrak, Celeste Banaag, David Burkhalter, Robert Cahalan, Songqing Chen, Lloyd Clark, Guido Colangeli, Corey Dakin, Hazem Eldakdoky, David Engelbrecht, John D. Evans, Daniel Fay, Bill Fink, Paul Fukuhara, Andrian Gardner, Pat Gary, Tom Giffen, Steve Glendening, Yonanthan Goitom, Lon Gowen, Sue Harzlet, Mohammed Hassan, Thomas Huang, Haisam Ido, Mahendran Kadannapalli, Ken Keiser, Paul Lang, Wenwen Li, Matthew Linton, Lynda Liptrap, Michael L. Little, Stephen Lowe, Richard Martin, Jeff Martz, Roy Mendelssohn, Dave Merrill, Lizhi Miao, Mark Miller, Nick Mistry, Matthew Morris, Doug Munchoney, Aruna Muppalla, Steve Naus, Slobodan Nickovic, Erik U. Noble, Robert Patt-Corner, Goran Pejanovic, Pete Pollerger, Chris Rusanowski, Todd Sanders, Gavin A. Schmidt, Alan Settell, Khawaja S. Shams, Bryan J. Smith, William Sprigg, Mike Stefanelli, Joe Stevens, Nicola Trocino, Tiffany Vance, Archie Warnock, Mike Whiting, Paul Wiese, Lisa Wolfisch, Huayi Wu, Yan Xu, and Abraham T. Zeyohannis.

The chapters were reviewed by external experts including Michael Peterson at the University of Nebraska at Omaha, Ian Truslove of the National Sea Ice Data Center, Chuanrong Zhang at the University of Connecticut, Stefan

Falke at Northrop Grumman, Xuan Shi at the University of Arkansas, Long Pham at NASA Goddard, Jian Chen at Louisiana University, Jin Xing at McGill University, Marshall Ma at Rensselaer Polytechnic Institute, Thomas Huang at NASA JPL, Chris Badure at Appalachia State University, Doug Nebert at FGDC, Rick Kim at the National University of Singapore, Rui Li at Wuhan University, and Alicia Jeffers at the State University of New York at Geneseo. Nanyin Zhou at George Mason University spent a significant amount of time formatting the book. Peter Lostritto at George Mason University proofed several chapters.

Thanks also go to CRC Press/Taylor & Francis acquiring editor Irma Britton, her assistant and production coordinator Arlene Kopeloff, Joselyn Banks-Kyle who helped to ensure that the manuscript was formatted according to standards adopted by CRC Press/Taylor & Francis. They provided insightful comments and were patient when working with us. Students who worked on the projects and participated in writing the chapters are greatly appreciated.

Finally, we would like to thank our family members for their tolerance and for bearing with us as we stole family time to finish the book.

Chaowei Yang would like to thank his wife Yan Xiang, his children Andrew Yang, Christopher X. Yang, and Hannah Yang.

Qunying Huang would like to thank her husband Yunfeng Jiang.

Zhenlong Li would like to thank his wife Weili Xiu and his son Mason J. Li.

Chen Xu's deepest gratitude goes to his beloved wife, Jianping Han, for her understanding, support, and patience.

Kai Liu would like to thank his wife, Huifen Wang.

Part I

Introduction to cloud computing for geosciences

Cloud computing is a new generation computing paradigm for sharing and pooling computing resources to handle the dynamic demands on computing resources posed by many 21st century challenges. This part introduces the foundation of cloud computing from several aspects: requirements from the domain of geoscience (Chapter 1), cloud computing concepts, architecture, and status (Chapter 2), and the technologies that enabled cloud computing (Chapter 3).

Chapter 1

Geoscience application challenges to computing infrastructures

Chaowei Yang and Chen Xu

This chapter introduces the need for a new computing infrastructure such as cloud computing to address several challenging issues including natural disasters, energy shortage, climate change, and sustainability in the 21st century.

1.1 CHALLENGES AND OPPORTUNITIES FOR GEOSCIENCE APPLICATIONS IN THE 21ST CENTURY

Geoscience is facing great challenges in dealing with many global or regional issues that greatly impact our daily lives. These challenges range from the history of the planet Earth to the quality of the air we breathe (NRC 2012a,b). This section examines the challenges of energy, emergency responses, climate, and sustainability.

1.1.1 Energy

Currently, about 80% of the world's energy demand is fulfilled by fossil fuels (IEA 2010). However, the reliance on fossil fuels is unsustainable due to two fundamental problems: first, fossil fuels are nonrenewable and eventually these resources will be depleted, and second, the consumption of fossil fuels is causing serious environmental and social problems, such as climate change and natural resource related conflicts. The U.S. Energy Information Administration (EIA) and the International Energy Agency (IEA) have predicted that global energy consumption will continue to increase by 2% every year, by which in 2040, the rate of energy consumption will have doubled the rate set in 2007 (IEA 2010). With limited achievements in developing more sustainable alternative energy, most of the predicted increasing energy consumption would come from fossil fuels. Hence, we are accelerating toward the depletion of fossil fuels and are producing more greenhouse gases. The objective to keep the average global temperature increase under 2 degrees Celsius above pre-industrial levels is becoming a bit too optimistic, as fundamental transformations in energy consumption are constantly evaded (Peters et al. 2012).

To achieve a secure energy future, IEA deems that the transparency of the global energy market ensured by energy data analyses and global collaboration on energy technology are crucial strategies to be taken. One implementation of advanced information-based energy consumption is through a smart grid, which uses digital technology for dynamically matching energy generation with user demand. Comprehensive data, collaborative sensors, and intelligent energy management have put an enormous challenge on advanced computing for developing capacities to embrace big data, supporting dynamic collaboration, and incorporating intelligence into the energy grid for smart energy consumption management (IEA 2010).

1.1.2 Emergency response

Natural and human-induced disasters are increasing in both frequency and severity in the 21st century because of climate change, increasing population, and infrastructure. For example, the 2003 avian influenza spread across all continents in just a few weeks from international imports and human transportation (Li et al. 2004). The flooding brought by hurricanes, tsunamis, and heavy rainfall cause the loss of tens of thousands of people each year around the world. Wildfires during drought seasons cause the loss of billions of dollars of assets. The leakage of hazardous materials, such as nuclear material and poisonous gas, also cause hundreds of people's lives each year (IAPII 2009). Responding to natural and human-induced disasters in a rapid fashion is a critical task to save lives and reduce the loss of assets.

Decision support for emergency response can only be best conducted when integrating a large amount of geospatial information in a timely fashion. For example, Figure 1.1 shows the flooding map of the 2005 New Orleans tragedy when Hurricane Katrina dumped water and swamped the entire city. The map shows the location and depth of the flooding ranging from 0 to 14 feet. The map could be a great decision support tool if it could be produced in a few hours after the hurricane for both the first responders and the residents of New Orleans for deciding whether to evacuate or not and where to conduct search and rescues. Unfortunately, it takes several weeks to collect all geospatial, meteorological, civil engineering, and other datasets to produce such a map (Curtis, Mills, and Leitner 2006). There are two difficulties for producing such a map in a few hours. First, the data are distributed across different agencies and companies and it takes a relatively long time to identify and integrate the datasets. Second, tens to hundreds of computers are needed for the simulations and flooding calculations to be completed in a few hours. Once the map is produced, the computing resources can be released. This demands an elastic computing

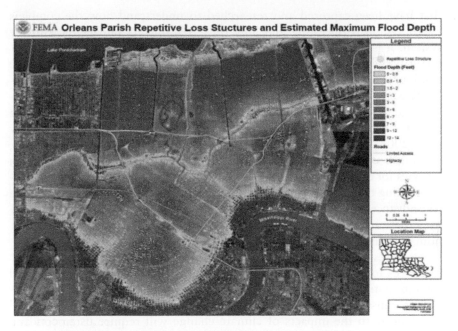

Figure 1.1 (See color insert.) Flooding depth map of New Orleans after Hurricane Katrina. (Courtesy of the Federal Emergency Management Agency [FEMA], Department of Homeland Security.)

infrastructure that can be allocated to process the geospatial data in a few minutes with or without little human intervention. Similar computing elasticity and rapid integrations are also required for other emergency responses, such as for wildfires, tsunamis, earthquakes, and even more so for poisonous and radioactive material emissions.

1.1.3 Climate change

Global climate change is one of the biggest challenges we are facing in the 21st century. Climate change requires scientific research to identify the factors that lead to answering how and why the climate is changing. The National Research Council recommends that further studies are needed to understand climate change for three different aspects: (1) advancing the understanding of climate change (NRC 2010a), (2) limiting the magnitude of future climate change (NRC 2011), and (3) adapting to the impacts of climate change (NRC 2010b).

Although the factors influencing climate change can be simply categorized into internal forcing (such as water and energy distribution) and external forcing (such as volcanic eruption, solar radiation, and human activities), in-depth understanding of climate change will require hundreds of parameters that have been captured and researched to understand how the complex Earth system is operating and how those parameters impact the climate system. It is a daunting task for scientists to build a variety of models to quantify the influence of the parameters and run numerous different model configurations and compare them with the observations to gain scientific knowledge. With limited computing resources, it becomes important to leverage the idle computing resources in a grid computing fashion, such as the UK climate prediction or the NASA Climate@Home projects (Stainforth et al. 2002). The management of the computing infrastructure will require disruptive storage, communication, processing, and other computing capabilityies for coordinating data and computing among the computers.

Limiting the magnitude of future climate change depends on the scientific advancements of climate change and at the same time, a well-conceived hypothetical scenario simulating the possibilities of climate change given human-induced key parameters, such as carbon dioxide and other greenhouse gases. This simulation requires a significant amount of computing resources to be ready for use in a relatively short time period to prepare for international and national negotiations; for example, supporting the carbon emission control decision making in a political process (IEA 2010).

Adapting to the impacts of climate change will require us to conduct many multiscale simulations (Henderson-Sellers and McGuffie 2012) including: (1) global scope simulation for supporting international policy-related decision making and (2) regional decision support with mesoscale climate model simulations based on, for example, sea level rise and coastal city mitigation. Higher-resolution simulations may be needed for property-related predictions because of climate change; for example, to support insurance policy making. Considering the climate impact in the near future, these simulations will become increasingly frequent and could be requested as a computing service when an insurance price quote for a house is requested. Each of the inquiries may invoke significant computing resources in a relatively short time period for real-time decision making, therefore presenting a spike requirement for the computing infrastructure.

1.1.4 Sustainable development

Sustainability, which emerged in the 1970s and 1980s (Kates and Clark 2000), benefits extensively from the advancement of computer science and information technology that provides toolsets for data collection, data management, computational modeling, and many other functionalities. Sustainability as a multidisciplinary study concerns complex interrelationships among the three areas of natural environment, economic vitality, and healthy communities (Millett and Estrin 2012).

The survival of human beings depends on the availability of freshwater on Earth. As the total amount of freshwater is only a small percentage of all water resources on Earth and its availability to individuals is being reduced due to the growing population and the shrinking freshwater supply, more regional conflicts are incurred by the shortage of water resources (Kukk and Deese 1996). Freshwater is an indispensable part in the production of many human commodity products. It enters the global circles of commodities, which complicates the planning for a sustainable use of water resources. The sustainable planning of water usage calls for a comprehensive modeling of water consumption at various scales from local to global. The process has been proved to be data intensive (Maier and Dandy 2000).

As the Earth becomes increasingly urbanized, it only creates more serious problems that challenge the overall health of a living environment of its dwellers. Such problems comprise urban sprawl, urban heat islands, sanitation-related health burdens, and pollution, as well as oversized energy and groundwater footprints. We are in urgent need of restrictions on urban sprawl, sanitizing of urban environments, and the building of more livable urban spaces (Gökçekus, Türker, and LaMoreaux 2011). Comprehensive urban planning is required. Such planning demands an enormous amount of data to be collected, processed, and integrated for decision making. The process would benefit from improving the availability of new computing resources.

The world population is predicted to be 9.3 billion by 2050 (Livi-Bacci 2012). With the enormous population to be sustained by an unsustainable consumption of fossil fuel-based energy and an increasingly smaller individual share of freshwater, with the growing number of populations living in gradually deteriorating urban environments, the sustainability of human society is in peril. In order to reverse this dangerous trend, systematic and bold policy changes need to be taken, which have to be backed by sound scientific research. As computational turns have been happening in various academic research domains, the research processes are being dramatically reshaped by digital technology (Franchette 2011).

Sustainability challenges often share characteristics of scale (e.g., sustainable energy consumption to be achieved locally, regionally, or globally) and heterogeneity (e.g., different factors contribute to freshwater availability and there are multiple solutions to the issue). The best solution has to optimize trade-offs among competing goals, which renders the process both data intensive and computing intensive. For example, we are at the early stage of an Internet of Things (Atzori, Iera, and Morabito 2010) characterized by devices that have sensors, actuators, and data processors built in. Such devices are capable of sensing, collecting, storing, and processing real-time data, for instance, from environmental monitoring, stock market records, or personal gadgets. The amount of data that have been collected is enormous. By the end of 2011, the number was 1.8 ZB, and is estimated to be 35 ZB by 2020 (Krishna 2011). The availability of big data has suggested a new paradigm of decision making driven by data. Computing innovations are needed to effectively process the big data.

1.2 THE NEEDS OF A NEW COMPUTING INFRASTRUCTURE

The challenges described in Section 1.1 call for a computing infrastructure that could help conduct relevant computing and data processing with the characteristics of enough computing capability, a minimized energy cost, a fast response to spike computing needs, and wide accessibility to the public when needed.

1.2.1 Providing enough computing power

Although computing hardware technologies, including a central processing unit (CPU), network, storage, RAM, and graphics processing unit (GPU), have been advanced greatly in past decades, many computing requirements for addressing scientific and application challenges go beyond existing computing capabilities. High performance computing (HPC) hosted by computing centers has been utilized for scientific research. Computing capabilities offered by HPC centers often fail to meet the increasing demands for such scientific simulations and real-time computing demands. Citizen computing is one approach that has been adopted by the scientific community to leverage citizen-idle computing cycles to address this problem. For example, the SETI@Home project utilizes citizen computers to help process signals from outer space to detect possible alien communications (Anderson et al. 2002). Because of the large volume of signals and possible algorithms to process those signals, billions of runs are needed through the combination. The requirement for computing resources will increase when more signals are picked by better sensors and more sophisticated algorithms are developed to better process the signals. Citizen computing becomes a good approach for this processing since there is no predictable time frame that we would like for finding aliens. Another example is the Climate@Home project, which utilizes citizen computers to help with running climate models for thousands of times to help climatologists improve predictions of climate models (Sun et al. 2012). This approach was also used to help biologists solve biological problems, such as Folding@Home (Beberg et al. 2009).

Another computing demand to address these challenges is to get people to help solve the problems. We often address this by shipping the problem to the public or crowdsourcing the problems. This type of emerging computing model helps problem solving, for example, to design urban transit plans or to validate patent application[1] by the public in a certain period of time. A prize is normally given to the best contributor and the intellectual property is owned by the problem submitter.

---

[1] See Peer to Patent at http://peertopatent.org/.

Both citizen computing and crowdsourcing are viable solutions for some 21st century challenges as described in the previous examples. However, neither has the proper guarantee of timeliness. Many challenges need to obtain computing power within a reasonable time frame. A new computing infrastructure is needed to support such challenges with timely requirements.

1.2.2 Responding in real time

In response to emergencies, most operational systems will require real-time delivery of information to support decision making and that one-second early warning or alert may help save more lives (Asimakopoulou and Bessis 2010). This real-time delivery of information also widely exists in other decision support environments, for example, in computing infrastructure coordination and scheduling. The real-time requirement demands fast allocation of large amounts of computing resources and fast release of the computing resources when the emergency response is completed.

Similarly, spiking access to computing resources is demanded by the Internet and social media, pushing the access to computing to the general public. Once a major event happens, such as the 2012 U.S. presidential election or a big earthquake, a significant amount of computing resources will be required in a relatively short time period for public response. This spike access may also exist with a spatiotemporal distribution. For example, around the globe, when the United States is in daytime, there will be spike access to computing infrastructure for public information or daily work. At the same time, Asia, on the other side of the Earth, will have the least access to their computing infrastructure and vice versa. How to utilize the computing power for this global spatiotemporal computing demand is also a challenge.

1.2.3 Saving energy

The past decades have seen a fast growth in processor speed. In contrast, the energy consumption for CPUs has also been reduced exponentially. The energy consumption of an entire computer system has also decreased significantly in past decades. In addition to the hardware part, the software part of computing management, job scheduling, and the management of hardware activity also moves toward saving energy to obtain better watts per giga (billions) floating point operations (GFLOP) and better watts per throughput of a computer. These achievements have been successfully utilized in a single computer and high performance computing (HPC) systems. However, the rapid increase in computing resources usage (personal computers, servers, and other intelligence terminals such as tablets and mobile devices) dramatically increases the global energy consumption. To reduce global energy consumption, especially for a distributed computing system which is required or demanded to deal with the 21st century challenges, we need a management system that can help pool and share the computing resources across geographically dispersed regions.

1.2.4 Saving the budget

Contemporary research or applications supported by computing normally have very disruptive computation usage. For example, when conducting scientific research or study in the academic environment, we may need to use a significant amount of computing resources in a relatively short time period, such as a half-day or three hours of lecture time out of the week. While the requirements of the computing resource for the demands are big, we do not use the resource on a continual basis or a less than 10% basis. Therefore, hosting the maximum number of private computing resources would either not be cost efficient, or buying such computing resources would have to be based on the utility model, or the so-called utility computing. A computing infrastructure would be ideal to charge only the portion used.

1.2.5 Improving accessibility

The popularization of the Internet and smart devices, such as tablets and smartphones, provides us with ubiquitous computing power that is comparable to earlier desktops. Many applications are deployed on smart devices and become available to the general public. For example, 85% of global phone owners are using their phones to send text messages, and 23% are using them to surf the Internet (Judge 2011). Most applications are also provided with smart device clients for the general public. This wide accessibility of computing from our daily smart devices provides the demand and opportunity to access distributed computing on a much broader basis other than the early networked desktops or laptops. And it becomes an ideal and natural requirement for accessing computing resources broadly by means of smart devices.

1.3 THE BIRTH OF CLOUD COMPUTING

1.3.1 Distributed computing

The ever-growing demand for computing power drives computer scientists and engineers to explore new paradigms to supply computing in a cost-effective way. Sharing computing resources dates back to the sharing of large-scale mainframes utilized by academia and companies in the 1950s, when the cost for a mainframe was extremely high (Richards and Lyn 1990).

In order to achieve the greatest return on the investment of a mainframe, strategies were developed to enable multiple terminals to connect to the mainframe and to operate by sharing CPU time. At that time, the sharing of computing resources was localized to where the mainframes were positioned. The sharing of mainframes has been considered to be the earliest model of cloud computing (Voas and Zhang 2009). The computing sharing model is very similar to the model that utility companies such as electric companies use to provide service to customers. John McCarthy, in a speech at MIT, first publicly envisioned computing to be provided as a general utility, a concept that later was thoroughly explored by Doulas Parkhill (1966). Herb Grosch (Ryan, Merchant, and Falvey 2011) even boldly predicted that the global requirements for computation could be fulfilled by 15 large data centers. The concept was adopted by major companies such as IBM to create a business model for sharing computing resources based on time-sharing technology.

In the 1990s, telecommunication companies crafted a new mechanism for improving the efficiency and security of remote data communication by creating a virtual private network (VPN) instead of the original physical point-to-point data connection. The VPN mechanism improved the effective utilization of network bandwidth. Thus, the concept of cloud computing was enlarged to incorporate the sharing of communication infrastructure, and the initially localized cloud computing model was able to support geographically dispersed users to benefit from the mainframes by leveraging the expanded Internet communication capacity.

The idea of cloud computing started at the early stage of networked computing and the name *cloud computing* comes from the general usage of the cloud symbol in the system diagrams of networked computing and communication systems (ThinkGrid 2013). Later, cloud referred to the Internet, which connects networks of computing systems through a communication infrastructure. The cloud computing concept represents a type of networked structure for managing and sharing computing resources, mainly for the purpose of maximizing a return on investment (ROI) on initial computing resources (Armbrust et al. 2010). Cloud computing in the modern IT environment expands CPU and bandwidth sharing, to share computing resources more thoroughly through hardware virtualization, service-oriented architecture, and delivering computing services as a type of utility through the Internet.

1.3.2 On-demand services

The service-based architecture enables computing services to be delivered on demand. Cloud computing is becoming a new paradigm of utility services, which provides computing power and applications as a service to consumers of either public or private organizations. Utility computing follows the commonly accepted business model of energy companies, which bases cost on actual usage, a pay-per-use business model. Hence, a consumer's computing demands can be satisfied in a more timely fashion by issuing new service requests to a provider, and the provider charges the consumer based on his/her actual usage.

1.3.3 Computing sharing and cost savings

Through the business model of utility computing, customers can be relieved of the burden of the constant investment of purchasing computing hardware and software, and save expenditure on system maintenance. By spending on the purchase of computing services, customers transfer the cost to the operations. Computing service providers assemble heterogeneous computing resources and allocate them dynamically and flexibly according to demands. By adding a middleware hardware virtualization to manage and broker the hardware, computing providers maximize their return on initial equipment through investments by reducing system idle time. Through cloud computing, consumers and providers form a win–win combination.

1.3.4 Reliability

Besides cost savings, applications on the cloud normally achieve improved reliability with shared redundancy either through expanding computing capacity of a single cloud or through integrating multiple clouds. Cloud computing providers can seamlessly add new hardware to the resource pool when more computing power is desired. Restricted by Service Level Agreements (SLAs), multiple services can be chained together. Cloud computing users hence are able to integrate multiple services from multiple providers, which potentially reduces the risks of exclusive reliance on a single provider and improves reliability. Studies have shown, by utilizing cloud computing, Web-based applications can improve their online availability (Armbrust et al. 2010).

1.3.5 The emergence of cloud computing

The advancements in computational technology and Internet communication technology help computing sharing go beyond simple time-based sharing at the physical component level (CPUs, RAM) to the virtualization-based sharing at the system level, which has been generally termed *cloud computing*. Once computing capability can be delivered as a service, it enables the external deployment of IT resources, such as servers, storage, or applications, and acquires them as services. The new computing paradigm allows adopters to promote the level of specialization for greater degrees of productive efficiency (Dillon et al. 2010). Cloud computing converges from the multiple computing technologies of hardware virtualization, service-oriented architecture, and utility computing, as will be detailed in Chapter 3.

Cloud computing frees consumers from purchasing their own computing hardware, software, and maintenance, and provides the following summarized features:

- Computing and infrastructure resources and applications are provided in a service manner.
- Services are offered by providers to customers in a pay-per-use fashion.
- Virtualization of computing resources enables on-demand provisioning and dynamic scalability of computing resources.
- The services are provided as integrated delivery including supporting infrastructure.
- Cloud computing is normally accessed through Web browsers or customized application programming interface (API).

Amazon was one of the first providers of cloud computing. Amazon Elastic Compute Cloud (EC2) abstracts generic hardware as Amazon virtual machines with various performance levels, which are then provided to customers as services that they can choose based on their needs. The *elasticity* of the services enables cloud consumers to scale the computing resource to match computational demands as required. For instance, current available EC2 computing power ranges from a small instance (the default instance) with a 32-bit platform (one virtual core and 1.7 Gbytes of memory and 160 Gbytes of storage) to various high-end configurations that are geared toward the most demanding tasks. For example, for high-throughput applications, the highest configuration is a 64-bit platform with 8 virtual cores, 68.4 Gbytes of memory, and 1690 Gbytes of storage.[1]

As cloud providers, Amazon launched the EC2 in 2006 and Microsoft started Azure in 2008. This was soon followed by many open-source cloud solutions becoming available, such as Eucalyptus. In addition to operating system level service, some cloud providers serve customers who are software developers with either support to all phases of software development or a platform for specialized usage such as content management. For example, Google enables developers to develop applications to be run on Google's infrastructure. More cloud computing providers are in the market of delivering software as a service. A classic example is Google Mail (Gmail).

---

[1] According to information provided by Amazon at http://aws.amazon.com/ec2/instance-types/.

1.4 THE ADVANTAGES AND DISADVANTAGES OF CLOUD COMPUTING FOR GEOSCIENCE APPLICATIONS

1.4.1 Advantages of cloud computing

The original concept and mechanism for sharing computing power were formed in the academic realm to enable as many users as possible to use mainframes simultaneously (Richards and Lyn 1990). Modern cloud computing provides academic researchers with computing power far beyond what they used to receive. In the age of the mainframe, the capacity of computing resources was restricted by the maximum capacity of the mainframe. When a requirement went beyond the existing capacity, new hardware had to be purchased. Because cloud computing brings the possibility of having computing demand being fully satisfied, academic computational researchers, who benefit from traditional mechanisms of shared computing facilities such as the supercomputing facilities, can leverage potentially unlimited computing resources (Dikaiakos et al. 2009). A savings in investments in purchasing new additional computing resources as well as costs by using more effective and powerful management can be realized (Xu 2012).

With the potential for accessing all computing resources in a virtualized computing environment, the operation of a critical scientific computing task can obtain increased reliability because failed computing resources would be replaced immediately by available resources through hardware, software, and geographic location redundancy. Standardized cloud computing APIs also allow cloud computing providers to supply their services seamlessly to multiple cloud computing brokers, and vice versa.

Computing intensive scientific applications have benefited from the computing power of supercomputing centers. Compared with the traditional supercomputing center, cloud computing platforms are more available and give users more control of their clusters. Rehr et al. (2010) demonstrate how Amazon EC2 can provide reliable high-performance support to general scientific computing. In summary, cloud computing provides a new capability for scientific computing with potentially unlimited availability of virtualized computing resources. Cloud computing is a new generation computing paradigm driven by the 21st century challenges that call for sharing and pooling of computing resources to satisfy dynamic computing demands (Yang et al. 2011).

1.4.2 Problems

The centralization of computing resources makes network infrastructure critical from end users to cloud computing facilities. If any of the critical infrastructures fail, the cloud services may not be available. A broad range of end users may lose access to their computing resources if the cloud computing facility is unavailable; for example, the 2010 Amazon Reston data center network outage caused a global impact to Amazon EC2 users.

The sharing of computing resources among organizations also causes the ultimate loss of control of computing resources by a consumer. For example, achieving security will become more complex and data concerning privacy may be difficult to put in a publicly available cloud computing environment. The problem is even worse when computing is shared across country boundaries.

Transmitting big data in the cloud or conducting computing intensive tasks may not be cost efficient when utilizing cloud computing. Therefore, cloud computing will not replace all other computing modes. For a while into the future, different computing modes will coexist (Mateescu, Gentzsch, and Ribbens 2011).

1.5 SUMMARY

This chapter introduces the demand for a new computing infrastructure from the standpoint of several 21st century challenging issues in Section 1.1. Section 1.2 analyzes the characteristics of such a new computing infrastructure. Section 1.3 discusses the evolution of distributed computing that led to the birth of cloud computing, and Section 1.4 introduces the benefits and problems of cloud computing for geoscience applications.

1.6 PROBLEMS

1. Enumerate three geoscience application challenges facing us in the 21st century and discuss the computing needs for addressing the challenges.
2. Enumerate two other challenges facing us in the geosciences or other domains and discuss the requirement for computing resources.
3. Natural disasters cost many human lives and a large amount of asset losses. In order to respond to the disasters and mitigate the impact, decision makers need timely and accurate information. Could you focus on one type of disaster and explore the information needed and produce the information in a timely fashion?
4. What are the common computing requirements for dealing with 21st century geoscience application challenges?
5. What is distributed computing? Could you enumerate four paradigms of distributed computing?
6. What is cloud computing and how does it relate to distributed computing?
7. What are the advantages and disadvantages of cloud computing?

REFERENCES

Anderson, D. P., J. Cobb, E. Korpela, M. Lebofsky, and D. Werthimer. 2002. SETI@ home: An experiment in public-resource computing. *Communications of the ACM* 45, no. 11: 56–61.

Armbrust, M., A. Fox, R. Griffith, A. D. Joseph, R. Katz, A. Konwinski, G. Lee et al. 2010. A view of cloud computing. *Communications of the ACM* 53, no. 4: 50–58.

Asimakopoulou, E. and N. Bessis. 2010. *Advanced ICTs for Disaster Management and Threat Detection: Collaborative and Distributed Frameworks*. Information Science Reference.

Atzori, L., A. Iera, and G. Morabito. 2010. The Internet of things: A survey. *Computer Networks* 54, no. 15: 2787–2805.

Beberg, A. L., D. L. Ensign, G. Jayachandran, S. Khaliq, and V. S. Pande. 2009. Folding@Home: Lessons from eight years of volunteer distributed computing. In *Parallel & Distributed Processing, IPDPS. IEEE International Symposium*, pp. 1–8. IEEE.

Curtis, A. J, J. W. Mills, and M. Leitner. 2006. Spatial confidentiality and GIS: Re-engineering mortality locations from published maps about Hurricane Katrina. *International Journal of Health Geographics* 5, no. 1: 44.

Dikaiakos, M. D., D. Katsaros, P. Mehra, G. Pallis, and A. Vakali. 2009. Cloud computing: Distributed Internet computing for IT and scientific research. *IEEE Internet Computing* 13, no. 5: 10–13.

Dillon, T., C. Wu, and E. Chang. 2010. Cloud computing: Issues and challenges. In *Advanced Information Networking and Applications (AINA), 24th IEEE International Conference*, pp. 27–33. University of York, UK, IEEE.

Franchette, F. 2011. Why is it necessary to build a physical model of hypercomputation? *Proceedings of AISB'11: Computing and Philosophy* 1: 97–104.

Gökçekus, H., U. Türker, and J. W. LaMoreaux, eds. 2011. *Survival and Sustainability: Environmental Concerns in the 21st Century*. Berlin, Heidelberg: Springer.

Henderson-Sellers, A. and K. McGuffie. 2012. *The Future of the World's Climate*. Oxford, UK: Elsevier.

IAPII (the InterAcademy Panel on International Issues). 2009. *Natural Disaster Mitigation: A Scientific and Practical Approach*, p. 146. Beijing: Science Press.

IEA. 2010. Energy Technology Perspectives: Scenarios & Strategies to 2050. *International Energy Agency*. https://www.iea.org/techno/etp/etp10/English. pdf (accessed March 18, 2013).

Judge, S. 2011. Texting, Social Networking Statistics Worldwide. http://www. mobilephonedevelopment.com/archives/1394 (accessed March 28, 2013).

Kates, R. W., W. C. Clark, R. Corell, J. M. Hall, et al. 2001. Sustainability science. *Science* 292: 641–642.

Krishna, A. 2011. Why Big Data? Why Now? *IBM Corporation*. http://almaden.ibm. com/colloquium/resources/Why%20Big%20Data%20Krishna.PDF (accessed March 18, 2013).

Kukk, C. L. and D. A. Deese. 1996. At the water's edge—Regional conflict and cooperation over fresh water. *UCLA J. Int'l L. & Foreign Aff.* 1: 21.

Li, J et al. 2004. Study on transmission model of avian influenza. *Information Acquisition, Proceedings. International Conference*. IEEE.

Livi-Bacci M. 2012. *A Concise History of World Population,* 5th ed, p. 271. Malden: MA: John Wiley & Sons.

Maier, H. R. and G. C. Dandy. 2000. Neural networks for the prediction and forecasting of water resources variables: A review of modelling issues and applications. *Environmental Modelling Software* 15, no. 1: 101–124.

Mateescu, G., W. Gentzsch, and C. J. Ribbens. 2011. Hybrid computing—Where HPC meets grid and cloud computing. *Future Generation Computer Systems* 27, no. 5: 440–453.

Millett, L. I. and D. L. Estrin, eds. 2012. *Computing Research for Sustainability.* Washington, DC: National Academies Press.

NRC. 2010a. *Advancing the Science of Climate Change,* p. 504. Washington, DC: National Academies Press.

NRC. 2010b. *Adapting to the Impacts of Climate Change,* p. 272. Washington, DC: National Academies Press.

NRC. 2011. *America's Climate Choices,* p. 118. Washington, DC: National Academies Press.

NRC. 2012a. *International Science in the National Interest at the U.S. Geological Survey,* p. 161. Washington, DC: National Academies Press.

NRC. 2012b. *New Research Opportunities in the Earth Sciences,* p. 117. Washington, DC: National Academies Press.

Parkhill, D. 1966. *The Challenge of the Computer Utility,* p. 207. New York: Addison-Wesley.

Peters, G. P., R. M. Andrew, T. Boden, J. G. Canadell, P. Ciais, C. L. Quere, G. Marland, M. R. Raupach, and C. Willson. 2012. The challenge to keep global warming below 2 [deg] C. *Nature Climate Change* 3: 4–6.

Rehr, J. J., F. D. Vila, J. P. Gardner, L. Svec, and M. Prange. 2010. Scientific computing in the cloud. *Computing in Science & Engineering* 12, no. 3: 34–43.

Richards, T. J. and R. Lyn. 1990. *Manual for Mainframe NUDIST: A Software System for Qualitative Data Analysis on Time–Sharing Computers.* Replee P/L.

Ryan, P., R. Merchant, and S. Falvey. 2011. Regulation of the Cloud in India. *Journal of Internet Law* 15, no. 4: 7.

Stainforth, D., J. Kettleborough, M. Allen, M. Collins, A. Heaps, and J. Murphy. 2002. Distributed computing for public-interest climate modeling research. *Computing in Science & Engineering* 4, no. 3: 82–89.

Sun, M., J. Li, C. Yang, G.A. Schmidt, M. Bambacus, R. Cahalan, Q. Huang, C. Xu, E. U. Noble, and Z. Li. 2012. A Web-based geovisual analytical system for climate studies in future. *Internet* 4, no. 4: 1069–1085.

ThinkGrid. 2013. Introduction to Cloud Computing. http://logicplus.com.au/MenuItems/Infrastructure/Downloads/Cloud-whitepaper.pdf (accessed March 18, 2013).

Voas, J. and J. Zhang. 2009. Cloud computing: New wine or just a new bottle? *IT Professional* 11, no. 2: 15–17.

Xu, X. 2012. From cloud computing to cloud manufacturing. *Robotics and Computer Integrated Manufacturing* 28, no. 1: 75–86.

Yang, C., M. Goodchild, Q. Huang, D. Nebert, R. Raskin, Y. Xu, M. Bambacus, and D. Fay. 2011. Spatial cloud computing: How can the geospatial sciences use and help shape cloud computing? *International Journal of Digital Earth* 4, no. 4: 305–329.

Chapter 2

Cloud computing architecture, concepts, and characteristics

Chaowei Yang and Qunying Huang

Cloud computing emerged as a result of the evolution of distributed computing (Armbrust et al. 2010; Yang et al. 2009) and as a response to the call for easy and fast sharing of computing resources to address challenges in the 21st century.

2.1 CONCEPTS

Accessibility to computing resources has been greatly enhanced since the early mode of distributed computing as time sharing of mainframe infrastructure. The sharing of mainframes can only be scheduled with a limited number of local terminals. The Internet relieves computing terminals from collocating with the mainframe, and high performance computing (HPC) increases the power of centralized computing resources. Grid computing further improves the sharing of computing resources in a plug and play fashion. Cloud computing provides virtually unlimited computing resources which can be easily accessed. With the advent of cloud computing, the frontier of computing resource sharing has been greatly advanced to utility-based, market-oriented computing.

The idea of cloud computing can be traced back to the 1950s; the conceptual model was formally proposed in the 1980s; and development started in the 1990s. Successful cloud services only became popular within the past several years (Voas and Zhang 2009). Cloud services started with serving e-mails and then expanded to include many other computing capabilities as services (e.g., computing power and storage) or the so-called everything as a service (XaaS) (Xu 2012; Yang et al. 2011). The industry offers many different types of cloud services ranging from the infrastructure level, such as Amazon Elastic Compute Cloud (Amazon EC2), to the application level, such as e-mail, document sharing, and doodle pooling.

Witnessing the heterogeneity of cloud services and providing guidance to the industry and agencies to offer or consume cloud services, the National Institute of Standards and Technology (NIST 2011) took leadership in defining the standards for cloud computing. Mell and Grance (2009) define cloud computing as "a model for enabling ubiquitous, convenient, and on-demand network access to a shared pool of configurable computing resources (e.g., networks, servers, storage, applications, and services) that can be rapidly provisioned and released with minimal management effort or service provider interaction." They further elaborate the definition to include five characteristics (on-demand self-service, broad network access, resource pooling, rapid elasticity, and measured service), three service models (software as a service, platform as a service, and infrastructure as a service), and four deployment models (private cloud, community cloud, public cloud, and hybrid cloud). The five characteristics defined by NIST are detailed in Section 2.2. They differentiate cloud computing from other distributed computing paradigms (Yang et al. 2011).

2.2 CLOUD COMPUTING ARCHITECTURE

Different cloud architectures are defined from the perspectives of cloud consumers, cloud providers, and cloud brokers. The variations of cloud computing architectures are also influenced by specific applications. The most popular architecture is the NIST general architecture (Liu et al. 2011) that serves as a vendor-independent conceptual architecture. NIST-defined architecture is adopted by this book and is used to introduce the concepts of cloud computing in this chapter.

Figure 2.1 is adapted from the NIST (Liu et al. 2011) reference architecture, which defines six different players in cloud computing. The cloud provider (e.g., Amazon) is the person or organization that maintains the computing hardware and software, and supplies cloud services to others.

The cloud auditor (e.g., security officer responsible for the *Federal Information Security Management Act, FISMA*) is the person or organization that monitors and assesses the security, performance, and privacy impact of cloud services provided by the cloud provider. The cloud consumer, such as an Amazon EC2 user, is the person or organization that procures and utilizes the cloud services. Similar to priceline.com brokering travel resources, a cloud broker is the person or organization that manages the use, performance, and delivery of cloud services, and negotiates the relationship between the cloud providers and cloud consumers. The cloud carrier is the intermediary that provides connectivity and transport of cloud services from cloud providers to cloud consumers. There are different business models for cloud computing. For example, a cloud consumer may procure cloud service from providers with or without going through the cloud broker. Users in this book are defined as the end users of domain applications running on the cloud services that are operated by cloud consumers.

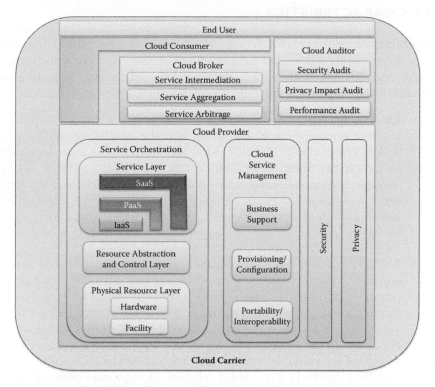

Figure 2.1 The cloud computing reference model (revised from the NIST cloud reference model). (From Liu et al. 2011.)

There are different service models provided by the cloud provider, such as Infrastructure as a Service (IaaS), Platform as a Service (PaaS), and Software as a Service (SaaS) as illustrated in Figure 2.1.

- *IaaS* provides the capability of provisioning computation, storage, networks, and other fundamental computing resources on which the operating systems and applications can be deployed. For example, through IaaS, a virtual server may be obtained and accessed, similar to accessing a traditional remote server. A popular representative of IaaS is Amazon EC2.
- *PaaS* provides a computing platform on which consumers can develop software using the tools and libraries from the cloud provider, and deploy the software onto cloud services. A typical example of PaaS is Windows Azure and Visual Studio provided by Microsoft.
- *SaaS* supports software and data to be hosted on the cloud and provided as services. Cloud consumers or users usually access this type of service through a Web browser. A popular application of SaaS is Gmail.

2.3 CHARACTERISTICS

Several characteristics of cloud computing (Mell and Grance 2009) differentiate it from other distributed computing paradigms (e.g., grid computing) (Yang et al. 2011). The differences among distributed computing paradigms are detailed in Chapter 3. An abbreviated description is provided below.

- *On-demand*—Cloud computing generally has a huge computing resource pool at the back-end to support the consumers' on-demand access. For example, Google's cloud service has more than 1 million servers across three dozen data centers worldwide (Newton 2010); Amazon,[1] Microsoft,[2] and other public cloud providers also run thousands to millions of physical servers. Private enterprise clouds also have hundreds of servers (Newton 2010). Therefore, cloud computing could give consumers unprecedented on-demand computation power backed by the large-scale computing pool.
- *Self-services with minimum interaction with the providers*—For the consumer, the capabilities of provisioning and releasing cloud computing resources often appear to be unlimited and can be purchased in any quantity at any time without interaction with the cloud providers, which is similar to our purchasing books from Amazon.com.
- *Broadband network and device independent access*—Cloud resources are available over the network and can be accessed through standard interfaces (e.g., mobile phones, laptops, and personal digital assistants [PDAs]). Data synchronization among heterogeneous devices involves complex operations. It is difficult to preserve and maintain the latest data and information copy among many different devices. Cloud computing can ease difficulties in data synchronization between different copies. Because data are synchronized automatically in the *cloud*, different devices can obtain the same version of data when connected to the Internet. For example, the Apple iCloud service enables users to synchronize software packages, e-mail addresses, apps, media files, and phone numbers among the iPhone, iPad, and Apple computers.
- *Resource pooling (for consolidating different types of computing resources)*—The provider's computing resources are pooled to serve multiple consumers using a multitenant model. Within this model, different physical and virtual resources are dynamically assigned and reassigned according to consumer demand (Liu et al. 2011). Traditional IT systems generally were using only 10 to 15% of their

---

[1] See ZDNet at http://www.zdnet.com/blog/open-source/amazon-ec2-cloud-is-made-up-of -almost-half-a-million-linux-servers/10620.

[2] See Neowin at http://www.neowin.net/news/main/09/11/02/inside-windows-azures-data -center-one-of-worlds-largest.

full capacity (Marston et al. 2011). Cloud computing significantly improves the sharing of computing resources across organizations and improves the utilization of computing resources up to 80% (Yang et al. 2011). At the same time, with the shared resource model based on virtualization, cloud computing reduces the cost for consumers to purchase, operate, and maintain the computing resources. From the cloud provider perspective, virtualization technologies can be applied to different levels of physical resources. Therefore, the latest physical resources can be added to the cloud resource pool, which reduces the cost of maintaining the and upgrading a cloud platform. On the other hand, cloud consumers are exempted from maintaining the infrastructure by outsourcing the responsibilities to cloud providers.

- *Rapid elasticity (for rapidly and elastically provisioning, allocating, and releasing computing resources)*—Within cloud services, applications can be configured to elastically acquire more resources to handle spiking workloads. When the loads decrease, the resources can be rapidly released within several seconds to minutes (Huang et al. 2010).
- *Measured service (to support pay-as-you-go pricing models)*—Cloud computing adopts flexible price models. Cloud resource usage can be monitored, controlled, reported, and charged, providing transparency of the consumed services for both cloud provider and consumer. Typically, cloud providers enable consumers to pay for computing usage by the hour without long-term commitment. Cloud providers also allow customers to reserve resources for a long term at a low, wholesale price. In addition, some providers (e.g., Amazon EC2) even provide the price bid (spot) option for bidding on unused cloud resources at an even lower price.
- *Reliability*—Cloud infrastructure consumers can get auxiliary services such as the continual improvement of proven physical infrastructures and data centers from providers. These services are in compliance with new policies, security patches, and upgraded without paying an additional cost. For example, the Amazon EC2 Service Level Agreement (SLA) commits 99.95% availability with all Amazon EC2 regions to consumers.

2.4 SERVICE MODELS

The service model is a categorization of the types of service available through cloud services. This book focuses on the most matured service models of IaaS, PaaS, SaaS, and DaaS, especially IaaS.

As illustrated in Figure 2.2, the layered computing logic framework includes the hardware, operating system, virtualization service (the core of cloud computing), platform management, software management, data

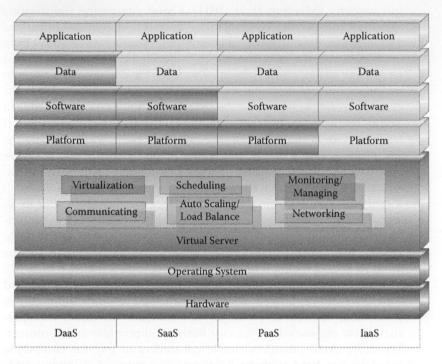

Figure 2.2 Cloud computing service model in the technical logic architecture. (Revised from Yang et al., 2011.)

management, and domain applications. If the cloud provider supplies the cloud service as a virtual computing server including the three lower layers, the service mode is called *IaaS*, such as Amazon EC2. If the cloud provider supplies the cloud service as a platform comprising the lower four layers for developing, testing, and deploying applications and software, the service mode is called *PaaS*, such as Windows Azure with Visual Studio. If the cloud provider supplies the cloud service as software including the lower five layers, the service mode is called *SaaS*, such as Gmail and Google Earth. If the cloud provider supplies the cloud service as a data service on top of all the layers except applications, then the service mode is called *DaaS*.

Applications could reside on any level of the services, IaaS, PaaS, SaaS, and DaaS, depending on the application functionalities. IaaS, PaaS, and SaaS have been provided as typical cloud services since the beginning of modern cloud computing. The big data challenge (The White House 2012) points out the need for DaaS (such as data.gov). The industry also provides other computing functions as a service, for example, Network as a Service (NaaS), Security as a Service (SECaaS), and Storage as a Service (STaaS). Collectively, all services models are labeled as everything as a service (XaaS) (Yang et al. 2011).

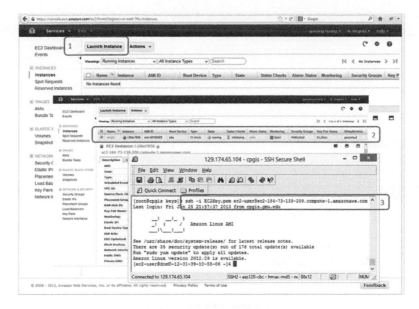

Figure 2.3 As labeled in sequence, cloud consumers can: (1) Log in to AWS and launch instance, (2) check status of instance, and (3) log in to the virtual instance to customize the server.

As a practical example, Figure 2.3 shows Amazon EC2, which is a representative of IaaS. The procedure starts with procuring a computer (virtual instance) by login to the EC2 console running in a Web browser and selecting the computer type to be started. After the instance is running, it can be accessed as a remote computer similar to the access to physical servers. The capability to deploy and test a Visual Studio project on the Azure cloud has been integrated with the Microsoft Visual Studio through minor configuration using an Azure account (Figure 2.4). Details of Amazon EC2 and Windows Azure usage are introduced in Chapter 4.

2.5 DEPLOYMENT MODELS AND CLOUD TYPES

Generally, there are four types of cloud deployments, including public cloud, private cloud, community cloud, and hybrid cloud (Liu et al. 2011) (Figure 2.5).

- *Public clouds* are available for open access and use by the general public. A public cloud system is usually provided, managed, and operated by a company. Consumers are charged for what they consume. Therefore, the public cloud is also known as a *commercial cloud*. So far, the public cloud is arguably the most popular and mature cloud type.

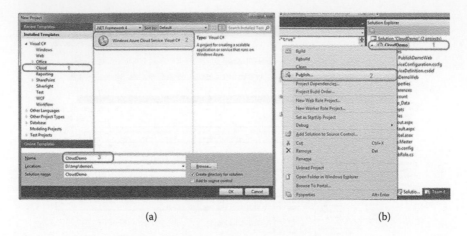

(a)   (b)

Figure 2.4 Create, develop, test (a), and deploy (b) applications onto Azure through Microsoft Visual Studio.

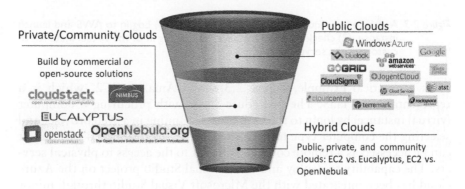

Figure 2.5 (See color insert.) Cloud types and software solutions.

- *Community clouds* are initiated to serve a community with common interests, such as mission, security, compliance, jurisdiction, and others. They are provided solely for a single organization or several organizations, and operated and managed internally or by a third party.
- *Private clouds* are provisioned for exclusive use by a single organization (e.g., business units). It may be owned, managed, and operated by a business, an academic institution, a government organization, or a consortium of them. These private clouds offer similar economic and operational advantages as the public clouds while allowing companies or organizations to retain absolute control over their IT resources. For example, Hewlett-Packard (HP) has made a shift to private cloud computing; the company consolidated its existing 85 data centers

with 19,000 employees to six cloud data centers with half the number of employees (Newton 2010). There are several open-source solutions to transform existing physical infrastructure into a private cloud or community cloud, such as Eucalyptus, CloudStack, OpenStack, and OpenNebula (Chapter 13) (Huang et al. 2012).

- *Hybrid clouds* are a combination of two or more of the aforementioned cloud types built for specific concerns or needs. For example, IT software enterprises may construct a hybrid cloud with two private cloud systems where one acts as the operational system, and the other serves development and testing. Also, public and private clouds can be combined to achieve cost-efficiency and on-demand computing power. For example, enStratus delivers brokerage for more than 10 cloud platforms[1]; OpenNebula can be used to build a private cloud, and its application programming interface (API) can be used to access Amazon EC2's public cloud.

2.6 REVIEW OF CLOUD COMPUTING RESOURCES

2.6.1 Commercial clouds

Enticed by the benefits of cost-efficiency, automation, scalability, and the flexibility of cloud services, many organizations are moving from traditional physical IT systems to cloud computing; commercial IT enterprises are increasingly providing cloud services in their products (Armbrust et al. 2010; Huang et al. 2013). For example, Amazon offers several types of cloud services including IaaS and PaaS. Windows Azure provides PaaS, with which users can quickly build, deploy, and manage applications across a global network of Microsoft-managed data centers. Details of Amazon Web Services and Windows Azure services are introduced in Chapter 4. This section briefly introduces several other popular commercial cloud services.

- *Google App Engine (GAE)*—Google outlined PaaS cloud computing in several papers from 2003 to 2006. The GAE was released as a prototype in 2008. Consumers can develop Web applications based on GAE APIs to be hosted in Google data centers. It offers automatic scaling for Web applications and dynamically allocates computing resources on-demand. GAE is cost-free up to a limited level of consumed resources. Fees are charged for additional storage, bandwidth, or instance operating hours required by the application.[2]
- *Salesforce*—The Salesforce platform provides different services for consumers to deliver transformative apps for customer relationship

---

[1] See Dell Software at http://enstratus.com/.
[2] See Google Developers at https://developers.google.com/appengine/.

management (CRM) business,[1] including: (1) Force.com to build employee apps and customize the Salesforce experience; (2) Heroku for creating customer apps in any programming language; (3) Database.com for sharing and storing data securely in the cloud database; (4) Site.com for delivering Web sites and digital marketing; (5) Identity for a single, social, trusted identity in the cloud; (6) AppExchange as a trusted cloud computing application marketplace; (7) ISVforce for building powerful commercial apps for cloud customers, and (8) Remedyforce for delivering next generation IT service management on the cloud.

- *Joyent*—Joyent is one of the IT vendors that provides cloud computing services including programmable load balancing, computing, storage, VPN (virtual private network), and Firewall (firewall) service. Different from many cloud providers (such as Amazon) who use Xen technology for virtualization, Joyent develops its own virtualization software, SmartOS, to ensure the stability and security of the service. SmartOS provides a combination of hardware and operating system (OS) virtualization to support efficient, reliable, and high-performing cloud computing.[2] Joyent has provided support for about 25% of the Facebook applications, and another one of its customers, LinkedIn, has more than 10 million visits within a single month. Another popular microblog service Identi.ca also selected Joyent as a platform.

- *Rackspace*—In July 2010, Rackspace, joined by leaders from across technology industries like NASA, Citrix, Dell, NTT DATA, Rightscale, and others, launched a new open-source cloud initiative OpenStack project.[3] From its initiation, OpenStack was designed to be open source and to build common ground for cloud providers as well as customers. OpenStack helped Rackspace to grow fast, offering an opportunity for Rackspace to compete with other big companies, such as Amazon. Currently, Rackspace cloud products are implemented on OpenStack.

- *GoGrid*[4]—GoGrid is a company owned by the U.S. Internet service provider ServePath, who has a long history in the hosting services and provides Internet hosting, dedicated servers, virtual hosts, and virtual servers.[5] GoGrid offers the world's first Web-based management console for multiserver management. Through the console or APIs, cloud consumers can provision and scale virtual and physical servers, storage, networking, load balancing, and firewalls in real time

---

[1] See Salesforce at http://www.salesforce.com/platform/overview/.
[2] See Joyent at http://joyent.com/.
[3] See OpenStack Cloud Software at http://openstack.org/.
[4] See GoGrid at http://www.gogrid.com/.
[5] See CrunchBase at http://www.crunchbase.com/company/servepath.

across multiple data centers within minutes. In addition, GoGrid also provides dedicated hosting services where cloud consumers can provision dedicated (physical) servers while integrating the flexibility and scalability of cloud server hosting services.

- *Verizon's CaaS*—In addition to telecommunications and additional services, Verizon Business also initiates cloud activities. For example, Verizon's CaaS allows customers to pay for data-center resources such as storage and application hosting dynamically based on the amount of resources they consume.
- *IBM*—SmartCloud is a cloud service that was released by IBM Corporation in June 2009. SmartCloud is capable of helping companies quickly create a cloud. It is a highly scalable cloud solution that combines infrastructure and platform capabilities to deliver elastic workload management, image life-cycle management, and resilient, high-scale provisioning on heterogeneous hypervisor and hardware platforms. Reservoir is an IBM and European Union joint research initiative for cloud computing that enables massive scale deployment and management of complex IT services across different administrative domains, IT platforms, and geographic regions.

2.6.2 Open-source cloud solutions

Several open-source programs have put forth efforts in investigating and developing technologies and infrastructure for cloud computing. For example, several open-source solutions are available for deploying and managing a large number of virtual machines to build a highly available and scalable cloud computing platform, such as Nimbus, Eucalyptus (Nurmi et al. 2009), OpenNebula (Sotomayor et al. 2009), CloudStack, and Virtual Workspaces. Details about Eculyptus, CloudStack, OpenNebula, and Nimbus are in Chapters 13 and 14. This section introduces several other open-source projects.

- *Amazon Elastic MapReduce (EMR)* is a Web service for efficiently processing and analyzing big data in the cloud. The Hadoop framework is integrated to implement a computational paradigm named *MapReduce* for distributed processing of large datasets on computed clusters, with Amazon EC2, Amazon S3, and other Amazon Web Services (AWS). EMR also supports Hadoop-related technologies such as Hive, Pig, and HBase.
- *Xen Cloud Platform* (XCP 2010) is a solution for cloud infrastructure virtualization. It uses Xen as the hypervisor, and provides management tools for virtual machines and networking on a single host. However, different from other open-source cloud solutions (e.g., Eucalyptus), XCP does not provide the overall architecture for IaaS. It is similar

to VMware's ESXi, which provides a tool to cope with the automatic creation, configuration, and maintenance of virtual machines.

- *TPlatform* (Peng et al. 2009) is a cloud solution that provides a development platform for Web mining applications. It is a typical PaaS inspired by Google cloud technologies. TPlatform's infrastructure is supported by three components: a scalable file system called *Tianwang File System* (TFS) similar to the Google File System (GFS), the BigTable data storage mechanism, and the MapReduce programming model.
- *Apache VCL* (VCL 2008) is an open-source system for remote accessing of computing resources through a Web interface. The resources are typically housed in a data center. VCL can serve as a broker to access stand-alone machines (e.g., a lab computer on a university campus). The VCL framework includes: (1) a Web client providing a user interface that enables the requesting and management of VCL resources; (2) a database server storing information about VCL reservations, access controls, and machine and environment inventory; and (3) multiple management nodes for controlling a subset of VCL resources.

2.7 SUMMARY

This chapter introduces the foundation of cloud computing based on NIST definitions. Section 2.1 introduces basic cloud computing concepts. Section 2.2 introduces cloud computing architecture. Section 2.3 introduces the characteristics of cloud computing. Section 2.4 introduces service models of cloud computing. Section 2.5 introduces cloud computing types based on a deployment model. Section 2.6 provides an overview of available cloud services and open-source solutions. The remainder of the book focuses on the technical side of how cloud computing can be utilized to support geoscience applications.

2.8 PROBLEMS

1. What is the NIST cloud computing definition?
2. What are the three most popular cloud computing service models?
3. What are the major characteristics of cloud computing? Based on your understanding, discuss which one is the most important.
4. What are the four different cloud computing deployment models and their user communities?
5. Enumerate four commercial cloud services and their vendors.
6. Enumerate four open-source cloud computing solutions and their owners.

REFERENCES

Armbrust, M., A. Fox, R. Griffith et al. 2010. A view of cloud computing. *Communications of the ACM* 53, no. 4: 50–58.

Huang, Q., J. Xia, C. Yang, K. Liu et al. 2012. An experimental study of open-source cloud platforms for dust storm forecasting. In *Proceedings of the 20th International Conference on Advances in Geographic Information Systems*, pp. 534–537. ACM.

Huang, Q., C. Yang, K. Benedict, S. Chen, A. Rezgui, and J. Xie. 2013. Enabling dust storm forecasting using cloud computing. *International Journal of Digital Earth*. doi:10.1080/17538947.2012.749949.

Huang, Q., C. Yang, D. Nebert, K. Liu, and H. Wu. 2010. Cloud computing for geosciences: Deployment of GEOSS Clearinghouse on Amazon's EC2. In *Proceedings of the ACM SIGSPATIAL International Workshop on High Performance and Distributed Geographic Information Systems*, pp. 35–38. ACM.

Liu, F., J. Tong, J. Mao, R. Bohn, J. Messina, L. Badger, and D. Leaf. 2011. NIST cloud computing reference architecture. *NIST Special Publication* 500, 292.

Marston, S., Z. Li, S. Bandyopadhyay, J. Zhang, and A. Ghalsasi. 2011. Cloud computing—The business perspective. *Decision Support Systems* 51, no. 1: 176–189.

Mell, P. and T. Grance. 2009. The NIST definition of cloud computing. *National Institute of Standards and Technology* 53, no. 6: 50.

Newton, J. 2010. Is cloud computing green computing? *GP Solo* 27, no 8: 28–31.

Nurmi, D., R. Wolski, C. Grzegorczyk, G. Obertelli, S. Soman, L. Youseff, and D. Zagorodnov. 2009 (May). The eucalyptus open-source cloud-computing system. *In Cluster Computing and the Grid. CCGRID'09. 9th IEEE/ACM International Symposium*, pp. 124-131. IEEE.

Peng, J., X. Zhang, Z. Lei, B. Zhang, W. Zhang, and Q. Li. 2009. Comparison of several cloud computing platforms. *Proceedings of 2nd International Symposium on Information Science and Engineering*, pp. 23–27.

Sotomayor, B., R. S. Montero, I. M. Llorente, and I. Foster. 2009. Virtual infrastructure management in private and hybrid clouds. *Internet Computing, IEEE* 13, no. 5: 14–22.

The White House, Office of Science and Technology Policy. 2012. Obama Administration Unveils "Big Data" Initiative: Announces $200 Million in New R&D Investments. http://www.whitehouse.gov/sites/default/files/microsites/ostp/big_data_press_release_final_2.pdf (accessed January 18, 2013).

VCL. 2008. Apache VCL. https://vcl.ncsu.edu/ (accessed January 18, 2013).

Voas, J. and J. Zhang. 2009. Cloud computing: New wine or just a new bottle? *IT Professional* 11, no. 2: 15–17.

XCP. 2010. Xen Cloud Platform. http://www.xen.org/products/cloudxen.html (accessed January 18, 2013).

Xu, X. 2012. From cloud computing to cloud manufacturing. *Robotics and Computer-Integrated Manufacturing* 28, no. 1: 75–86.

Yang, C., M. Goodchild, Q. Huang et al. 2011. Spatial cloud computing: How can the geospatial sciences use and help shape cloud computing? *International Journal of Digital Earth* 4, no. 4: 305–329.

Yang, C. and R. Raskin. 2009. Introduction to distributed geographic information processing research. *International Journal of Geographical Information Science* 23, no. 5: 553–560.

Chapter 3

Enabling technologies

Zhenlong Li, Qunying Huang, and Zhipeng Gui

Cloud computing provides (1) virtually unlimited storage to archive and manage the increasing amount of data, (2) on-demand computing capability to support computing intensive applications, and (3) the services to effectively utilize the network, storage, and computing resources. These capabilities are enabled by a variety of technologies that contributed to the emergence of cloud computing. To help readers understand how cloud computing emerged and evolved, this chapter introduces the key enabling technologies for cloud computing, including hardware advancements, distributed computing, virtualization, distributed file systems, and Web x.0.

3.1 HARDWARE ADVANCEMENTS

3.1.1 Multicore and many-core technologies

High performance computing (HPC) has become an indispensable tool for efficiently solving complex science, engineering, and business problems. However, the single-core and multithread computing model used by traditional HPC is unable to meet intensive computing demand (Pase and Eckl 2005). To fill this gap, multicore technology emerged and has been used for cluster computing since the late 1990s. Multicore technology enables multiple central processing unit (CPU) cores to be assembled on a single chip to share cache and memory. Through balancing the tasks among multiple CPU cores, the overall performance of multicore systems can be improved dramatically. Similarly, many-core technology enables tens to hundreds of processing cores to be integrated in a chip, such as the graphics processing unit (GPU), which further improves the computation power. Multicore and many-core technologies are widely adopted in HPC environments (Chai, Gao, and Panda 2007) due to the characteristics of low electricity consumption, efficient space utilization, and favorable performance in shared memory operations (Sagona 2009).

How does cloud computing benefit from multicore and many-core technologies? In the cloud computing era, cloud providers typically build multiple distributed data centers with massive physical computers to support consumers provisioning on-demand cloud resources. Each physical computer can be shared and accessed by multiple consumers simultaneously, and the access requests are assigned to different CPU cores. With the Infrastructure as a Service (IaaS) model, each consumer can utilize the computing power of the physical computer by means of the virtual machine (Section 3.3). Under such circumstances, physical computers in the cloud data centers are involved in heavy-duty computations and throughput. The parallel processing capabilities of multicore and many-core processors help relieve the processing burden of each physical computer. Therefore, the multicore and many-core technologies enable cloud providers to build energy-efficient and high-performance data centers that are the fundamental building blocks of cloud computing.

3.1.2 Networking

Cloud computing is a model in which resources and services are abstracted from the underlying infrastructure and provide on-demand and elastic services in a multitenant environment. Networking is critical to cloud computing in at least two aspects:

- The cloud provider delivers services to cloud consumers and users through the network. The computer network connects all the cloud components together. Without networks, consumers cannot access their cloud services. Computing infrastructures, software platforms, applications, and data can only be physically accessed by consumers and users.
- The networking relationship between cloud consumers and providers defines the technological variance among different deployment models. In a private cloud, consumers and providers are within the same trusted network boundary. In a public cloud, all cloud services are provisioned through the cloud carrier's network. In a hybrid cloud, a secured connection may exist between the consumer's network and the provider's network. In a community cloud, the network connection structure depends on the architecture of the organization operating the cloud.

3.1.3 Storage

Storage is the device that is traditionally referred to as the hard disk, tape system, and other devices connected to a computer. Cloud storage refers to the storage that can be provided to consumers as a service including network hardware, online storage, online backup, online documents, and others. As an essential part of a cloud service, cloud storage is a virtualized extensible resource.

In the cloud computing environment, it is still very challenging to store and manage massive data. A complex cloud service usually needs a large volume of storage to ensure elastic support for consumers. Fortunately, the fast developing hard-drive technologies have resulted in a constant increase of storage capacity as well as a price decrease. Cloud storage is an important component of the cloud computing infrastructure. It will have a significant impact on the development of cloud services in the future.

3.1.4 Smart devices

Cloud service offers consumers the ability to access information on-demand from a variety of devices independent of the hardware. The proliferation of various smart devices (e.g., smartphones and tablet computers) accelerates the development of cloud service by enriching its access channels for cloud consumers and users. For example, when traveling in a car, querying online maps will quickly obtain a route based on real-time traffic conditions through a traffic cloud service. With the capabilities of using different wireless networks (e.g., WiFi, 3G, and 4G), smart devices help significantly popularize cloud services.

3.2 COMPUTING TECHNOLOGIES

3.2.1 Distributed computing paradigm

The idea of distributed computing is to utilize computing capacity from multiple distributed computers to address larger computational problems. With distributed computing, a large-scale task is divided into many subtasks, each of which is solved by one or more computers. Distributed computing includes several variants such as cluster computing, grid computing, and utility computing.

Even though controversies still exist with regard to the relationships between cloud computing and distributed computing variants, it is generally accepted that distributed computing is the most obvious predecessor technology that enabled the inception of cloud computing (Youseff, Butrico, and Silva 2008), and that the initial idea of cloud computing evolved from distributed computing. This evolution has been a result of the shifting focus from an infrastructure that delivers storage and computing resources (such as the case in grid computing) to one that is economically based (Foster et al. 2008). Cloud computing aims to make extensive and highly virtualized resources easily available and scalable over the Internet as a service. These resources are specialized to various services including software, platform, and infrastructure.

3.2.2 Computing architecture model

Two models that emerged with the evolution of distributed computing paradigms were the client/server (C/S) model and the browser/server (B/S) model.

- The C/S model was introduced in the 1980s and offered a central repository of computers while personal computers and workstations served as terminals, allowing individuals to run programs locally (Tograph and Morgens 2008). C/S describes the relationship between two collaborating computer programs where one sends a request from the client, and another one on the server handles the request.
- The B/S model emerged with the development of the Internet and Web technologies, and a Web browser was used as the client program. This model unifies the different client programs in the C/S model, and simplifies the system development, maintenance, and usage.

Cloud computing enables convenient, on-demand network access to a shared pool of configurable computing resources, and cloud services are accessible from various client devices through a thin client interface, mostly Web browsers (Mell and Grance 2011). From this point of view, B/S is the key computing architecture model in the evolution of cloud computing.

3.3 VIRTUALIZATION

One of the core features of cloud computing is the abstraction of the physical implementation to hide technical details from consumers (e.g., which hardware is actually running the application, where it is located, and other configuration details). This functionality is supported by virtualization technology. Virtualization supports a cloud computing infrastructure (e.g., Amazon EC2) to provide computation capacity to consumers from remote locations through the Internet (Liu and Orban 2008). Virtualization also enables a computing system to dynamically acquire or release computing resources so that the system is resilient to component failures. Server consolidation, runtime guest migration, and security against malicious code are a few of the most compelling reasons for the use of virtualization (Liu and Orban 2008). With virtualization, cloud computing is able to create a dynamic number of virtual machines (VMs) depending on the needs of the application. For example, depending on the computing intensity and nature, a task may need the computing resource of a single CPU to hundreds of CPUs. Virtualization enables the scalability and flexibility of cloud computing by hiding the deployment details from the consumers (Vaquero et al. 2008). For example, with virtualization,

IaaS are able to split, assign, and dynamically resize computing resources to create ad hoc systems on demand.

3.3.1 Virtualization implementation

Hypervisor, which establishes an abstraction layer between the VMs and the underlying hardware, is a program that implements virtualization and allows multiple operating systems to share a single hardware host. Hypervisor can capture CPU instructions, and act as a coordinator for instructions to access hardware controllers and peripherals (Figure 3.1).

A wide range of virtualization approaches exists. Among them, three leading approaches are full virtualization, paravirtualization, and hardware virtualization (Sahoo, Mohapatra, and Lath 2010).

- *Full virtualization*—This is based on the host/guest paradigm where each guest runs on a virtual imitation of the hardware layer. In a full virtualization environment, hypervisor is running on the hardware directly, and the guest operating system (OS) running VM is managed by the hypervisor. The guest OS has no knowledge of the host's OS because it is not aware that it is not running directly on real hardware.
- *Paravirtualization*—The main drawback of a full virtualization approach is the performance overhead introduced by the hypervisor. One way to alleviate this overhead is to change the OS of VMs and let VMs know that they are running in a virtual environment and able to work together with the hypervisor. This method is called *paravirtualization*. This approach presents each VM with an abstraction of the hardware that is similar to but not identical to the underlying physical hardware. Paravirtualization attempts to provide most functions directly from the underlying hardware instead of abstracting it.
- *Hardware virtualization*—This is a virtualization solution where the hypervisor is embedded in the circuits of a hardware component instead of being supported by a third-party software application.

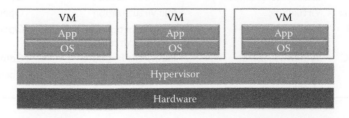

Figure 3.1 Implementation of virtualization.

3.3.2 Virtualization solutions

Currently, a wide range of virtualization solutions have been developed including both commercial and open-source solutions. For example, Xen, VMware, and VirtualBox support a wide spectrum of operating systems as either hosts or guests. As a result, the software compatibility issue is dismissed by obtaining the homogenization at the software level (McEvoy and Schulze 2008). Table 3.1 shows the widely used virtualization solutions, related cloud services, and their virtualization approaches.

- *Xen*—This is an open-source solution utilizing paravirtualization technology (Barham et al. 2003). To use Xen, the core codes of the physical machine's OS need to be modified. Therefore, Xen is suitable for open-source OS such as BSD, Linux, and Solaris, but is not suitable for proprietary OS, such as Windows, as their code-bases are protected. Xen is being widely adopted by Linux hosting providers, like Amazon EC2, GoGrid, 21Vianet CloudEx, and Rackspace Mosso.
- *KVM* (Kernel-based Virtual Machine)—This is a full virtualization solution for Linux on x86 hardware containing virtualization extensions.[1] Many cloud solutions (e.g., Nimbus, CloudStack, and OpenNebula) support both Xen and KVM virtualization.
- *VirtualBox*—This is another open-source solution under the terms of the GNU General Public License (GPL), Version 2. It is an enterprise level x86 and AMD64/Intel64 virtualization product. VirtualBox runs on Windows, Linux, Macintosh, and Solaris hosts and supports a large number of guest OSs including but not limited to

Table 3.1 Virtualization Solutions and Related Cloud Services

Hypervisor	Cloud Services	Virtualization Approach	Software Type
Xen	Amazon EC2, GoGrid, 21Vianet, CloudEx, RackSpace Mosso	Paravirtualization	Open Source
KVM	CloudStack, Nimbus	Full virtualization	Open Source
VirtualBox	VirtualBox	Full virtualization	Open Source
SmartOS	Joyent	A combination of hardware and OS virtualization	Open Source
Virtual PC	Windows Azure	Paravirtualization	Commercial
VMware	AT&T Synaptic, Verizon CaaS	Full virtualization	Commercial

---

[1] See KVM at http://www.linux-kvm.org/page/Main_Page.

Windows (NT 4.0, 2000, XP, Server 2003, Vista, Windows7), Linux (2.4 and 2.6), and OpenBSD.[1]

- *SmartOS*—This is also an open-source specialized hypervisor platform based on Illumos.[2] It supports two types of virtualization, OS Virtual Machines and KVM Virtual Machines. OS Virtual Machines is a lightweight virtualization solution offering good performance and all the features Illumos has. KVM Virtual Machines is a full virtualization solution for running a variety of guest OSs including Linux, Windows, BSD, and more.[3]
- *VMWare*[4]—This is a commercial product using a full virtualization solution. VMware VMotion is capable of moving a running VM instantaneously from one host to another. Its performance is faster than some other editions of VMware technology, and it is deployed in production by 70% of VMware customers, and leverages the complete virtualization of servers, storage, and networking. VMware VMotion uses VMware's cluster file system to control access to VM's storage. Within a VMotion, the active memory and precise execution state of a VM is rapidly transmitted over a high-speed network from one physical server to another, and access to the VMs disk storage is instantly switched to the new physical host. Since the network is also virtualized by VMware ESX, the VM retains its network identity and connections, ensuring a seamless migration process.

3.4 DISTRIBUTED FILE SYSTEM

Even though the fast, well-developed data storage devices (e.g., hard disk drives) have greatly increased massive data storage capacities as well as data read/write speed, the failure of disk hardware can create a single point of system failure resulting in data loss. Since the storage hardware cannot guarantee high reliability, it became a strategy for many big IT companies to build a reliable distributed data storage system with a group of cheap hardware to provide the high availability, reliability, and scalability for the data storage service. Such a storage system is based on a specialized file system called a *distributed file system* (DFS).

Cloud storage service is an essential capacity of cloud service, which provides virtually unlimited, secured and fault-tolerant data storage and access functions. Many popular cloud storage services such as Dropbox, iCloud, Google Drive, SkyDrive, and SugarSync have been enthusiastically

[1] See VirtualBox at https://www.virtualbox.org/.
[2] See Illumos at http://wiki.illumos.org/display/illumos/illumos+Home.
[3] See SmartOS at http://smartos.org/.
[4] See VMware at http://www.vmware.com/products/datacenter-virtualization/vsphere/vmotion.html.

used by public consumers. Availability, reliability, and scalability offered by DFS are of particular importance for the cloud storage service. Therefore, cloud services utilize the distributed data storage technology to store data.

The following subsections introduce the basic ideas and characteristics of DFS and the two most popular DFS implementations, the Google file system and the Hadoop Distributed File System.

3.4.1 Introduction to the distributed file system

A file system is an abstraction to store, retrieve, and update a set of files. It manages the access to the data and the metadata of the files, and the available space of the device(s), which contain the files. According to storage medium and usage purposes, file systems can be classified into disk/tape file systems, distributed file systems, and some other special-purpose file systems.[1]

The distributed file system (DFS) is a type of file system that allows access to files from multiple hosts, via a computer network (Yeager 2003). This makes it possible for multiple machines to share files and storage resources.[2] DFS usually serves on top of the local file systems (lower-level file system). The client nodes do not have direct access to the underlying block storage but interact over the network using a protocol. It can restrict access to the file system depending on access authorization lists or capabilities on both the servers and the clients, and on how the protocol is designed. In contrast, in a shared disk file system, all nodes have equal access to the block storage where the file system is located. On these systems, the access control must reside on the client. DFSs may include facilities for transparent replication and fault tolerance.

With the advancement of distributed computing and cloud computing, many DFSs are proposed and developed to access large scale and big data application and to meet other cloud service requirements, such as Amazon S3 and Oracle Automatic Storage Management Cluster File System (Oracle ACFS) (Shakian 2011).

3.4.2 Google File System

The Google File System (GFS) (Ghemawat, Gobioff, and Leung 2003) is a scalable distributed file system developed by Google for large distributed data-intensive applications. It provides fault tolerance while running on inexpensive commodity hardware.[3]

---

[1] See File System at http://en.wikipedia.org/wiki/File_system.
[2] See Clustered File System at http://en.wikipedia.org/wiki/Distributed_file_system.
[3] See Google File System at http://en.wikipedia.org/wiki/Google_File_System.

While sharing many of the same goals as previous distributed file systems, GFS has explored different designs by observing and analyzing Google's application workloads and technological environment. By comparing it with the Network File System (NFS) (a widely adopted traditional DFS for Linux and UNIX, which was developed by Sun Microsystems in 1984), the major differences are as follows (Ghemawat, Gobioff, and Leung 2003):

- *Bigger chunks*—Google chunk size is 64MB. This design is more suitable for big files and then provides better I/O performance.
- *Multiple data replication*—To guarantee the reliability, GFS provides at least three data redundancies.
- *Not local caching*—Unlike NFS, which offers local caching at the client side, GFS does not support local caching.

By adopting the above design, GFS can deliver high aggregate performance to a large number of clients, and it can offer very high scalability, reliability, and availability. GFS is not only widely deployed within Google as the storage platform for generating, processing, and archiving data, but also for research and development efforts that require large datasets.

3.4.3 Apache Hadoop Distributed File System

The Hadoop Distributed File System (HDFS) (Borthakur 2007; Shvachko et al. 2010) is a distributed file system designed to run on commodity hardware. HDFS serves as one important module (and also a subproject) of the Apache Hadoop project, which is a widely used open-source software framework for reliable and scalable distributed computing. Many IT companies such as Yahoo, Intel, and IBM have adopted HDFS as the big data storage technologies.

Just like GFS, HDFS is designed with the following common characteristics[1]: (1) Failures are normal rather than the exception; (2) Large datasets (huge file size, typical in GB to TB); (3) Streaming data access, for example, the file is sequentially read; (4) Append to write; (5) Hundreds of concurrent write accesses; (6) Bandwidth is more important than latency, and (7) Moving computation is cheaper than moving data. This is especially true when the size of the dataset is large. Therefore, HDFS is highly fault-tolerant and is designed to be deployed on low-cost hardware. HDFS provides high-throughput access to application data and is suitable for applications that have large datasets. It relaxes a few POSIX requirements to enable streaming access to file system data.

Both GFS and HDFS adopt master/slave architecture. A master server (NameNode) manages the file system namespace and regulates access to

---

[1] See Hadoop at http://hadoop.apache.org/docs/stable/hdfs_design.html.

files by clients. A number of slaves (DataNodes), usually one per node in the cluster, manage storage attached to the nodes that they run on. The DataNodes are responsible for serving read and writing requests from the file system's clients. The DataNodes also perform block creation, deletion, and replication upon instruction from the NameNode.

Advanced distributed file system technologies can provide high availability, reliability, and meet the requirement for large-scale and high-performance concurrent access. It has become a foundation of cloud computing for big data storage and management, and is also utilized in conjunction with parallel data processing frameworks, such as Map Reduce.

3.5 WEB X.0

The World Wide Web (WWW or *the Web*) is an information-sharing model that connects distributed, multiple sources and heterogeneous information using hypertext transport protocol (HTTP). The WWW provides a massive system of systems to allow Web consumers access to vast amounts of information over a networking infrastructure—the Internet.

- *Web 1.0*—The first generation emerged as early as 1993 when the hyperlinks among Web pages began with the release of the WWW to the public.[1] The primary focus during this stage was to build the Web and make it public and commercial. The Web pages were static and consumers could only view them, rather than contribute to the content.[2]
- *Web 2.0*—The phrase *Web 2.0* was first coined in 2004 by O'Reilly Media,[3] referring to a second generation of the Web enabling consumers to interact, collaborate, and coordinate with each other. Different from Web 1.0, Web 2.0 enabled users to do more than just view Web pages by providing dynamic editable Web pages, such as blogs, wikis, and various Web applications all through their Web browsers. This capability distinguishes Web 2.0 from a general information sharing model; it evolved into a new computing paradigm: "network as platform" computing,[4] which serves as one of the fundamental enabling technologies for cloud computing.
- *Web 3.0*—This is the Web's third generation aiming to make the Web more intelligent by adapting the technologies of semantic Web, data mining, machine learning, and artificial intelligence. Therefore,

---

[1] See Tim Berners-Lee at http://en.wikipedia.org/wiki/Tim_Berners-Lee.
[2] See Web 1.0 at http://en.wikipedia.org/wiki/Web_1.0#cite_note-3.
[3] See Paul Graham at http://www.paulgraham.com/Web20.html.
[4] See O'Reilly at http://oreilly.com/Web2/archive/what-is-Web-20.html.

Web 3.0 is also called the *Intelligence Web*.[1] At the time of this writing, no real Web 3.0-based service is available yet. Here is an example providing the basic idea of what a Web 3.0 service looks like. If someone wants to go out for dinner and watch the Avatar movie, with Web 2.0 he/she might have to search the movie to find out the theater near his/her location and then search the restaurants near the theater. With Web 3.0, he/she just needs to search "Avatar movie and dinner," and the search engine will do all the remaining work in a cloud computing environment and provide a list of options considering the locations for the theater and restaurant sorted by available time slots and prices.

- *Web 4.0*—This is still a developing concept with no clear and exact definition formed. It is generally believed that in the Web 4.0 era, computers will be much cleverer empowered by more intelligent human–computer interactive technologies, such as mind-controlled interfaces (Aghaei, Nematbakhsh, and Farsani 2012). For example, if someone wants to search today's weather information, he/she might only need to sit in front of the computer and think of the desired information in his/her mind, and the Web 4.0-based service will read the person's mind and respond.

The following subsections introduce several important technologies/ characteristics associated with Web x.0 that serve as key enabling technologies for cloud computing.

3.5.1 Web services

A Web service is a software system designed to support interoperable machine-to-machine interaction over a network.[2] Web Services Description Language (WSDL) is a machine and human-readable format designed for describing the Web service interface based on EXtensible Markup Language (XML). Simple Object Access Protocol (SOAP) is a protocol specification that enables other systems to interact with Web services through exchanging structured information over the Web using HTTP. Even though WSDL and SOAP are developed to describe the behavior of a standard Web service, they are not indispensable. Over the past few years, many Web 2.0 applications have switched from SOAP-based Web services to REST-based (Representational State Transfer) communications (Benslimane, Dustdar, and Sheth 2008), which has emerged as the predominant Web service design model. REST-based services allow consumers to retrieve information through simple HTTP methods such

---

[1] See Lifeboat Foundation at http://lifeboat.com/ex/Web.3.0.
[2] See W3C at http://www.w3.org.

as GET, POST, PUT, and DELETE. The generic Web-based APIs (application programming interfaces) such as Google APIs and Yahoo APIs are good examples of REST-based services. In the geospatial domain, the standards of Web Mapping Service (WMS), Web Coverage Service (WCS), and Catalogue Service for the Web (CSW) developed by the Open Geospatial Consortium (OGC)[1] are also based on REST style.

The Web is designed for improving information accessibility, and the wide usage of Web services, both SOAP-based and REST-based, has significantly enhanced this ability. Furthermore, Web services provide a standard means of interoperation between different software applications, running on a variety of platforms and/or frameworks,[2] which greatly advances the information interoperability over the Web.

An important characteristic of cloud computing is the *services over the Internet*, featured by the X as a service (XaaS) architecture. Web services, the building blocks for various Web applications, are playing an important role in the evolution of cloud computing, especially in the SaaS and DaaS tiers.

3.5.2 Service-oriented architecture

In software engineering, a service-oriented architecture (SOA) is a service-based component model including a set of principles and methodologies for designing and developing software in the form of interoperable services.[3] An SOA-based system is comprised of many loosely coupled services. Each service is a functional unit of the system and is defined by the service interface. This interface definition is independent of programming languages, operating systems, and hardware. The implemented services are published over a network (e.g., the Web) in which other developers can access and reuse them in other applications. Three major benefits of using SOA are component reusing, existing system integration, and language/platform independence.

- *Component reusing*—Since each component is defined and implemented as a Web service, these services can then be published and accessed by other users over the Web. To further improve the reuse capability, a set of Web service standards are developed, such as REST, SOAP, and WSDL.
- *Existing system integration*—SOA can be used to integrate existing systems by defining the Web interface of the system functions and

---

[1] See OGC at http://www.opengeospatial.org/.
[2] See Web Services Architecture at http://www.w3.org/TR/ws-arch/.
[3] See Service-Oriented Architecture at http://en.wikipedia.org/wiki/Service-oriented_architecture.

making them available to the enterprise in a standard agreed-upon way (Geoffrey 2009). These services are implemented by connecting to the existing functionalities and accessed to build a new system. Under such a scenario, SOA is acting as the application glue for sticking the existing functions together.

- *Language and platform independent*—Most contemporary Web service standards are based on XML (eXtensible Markup Language), a platform-independent markup language that is both human-readable and machine-readable. The Web service behaviors of input, output, and processing are well defined by XML-based standards such as WSDL, providing a common language-independent mechanism for implementing and consuming these services. When these services are published through SOAP to the Web, they can be accessed via any computer connected to the Internet, which is independent of the platform.

Cloud computing, to a large extent, leverages the concept of SOA, especially in the SaaS and PaaS layers. Though there are some overlaps between the concepts of cloud computing and SOA, they have a different emphasis. SOA is an architecture focused on answering the question of "how to develop applications" while cloud computing is an infrastructure emphasizing the solution of "how to deliver applications."[1] Figure 3.2 depicts how the Web and SOA technologies have helped the evolution of cloud computing.

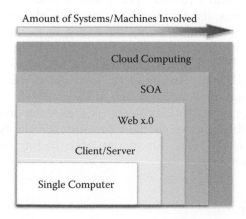

Figure 3.2 The evolution of cloud computing.

---

[1] Systems Engineering at MITRE at http://www.mitre.org/work/tech_papers/tech_papers_09/09_0743/09_0743.pdf.

3.6 CONCLUSION

The emergence, evolution, and maturity of cloud computing is driven and enabled by a variety of technologies as discussed in this chapter. The multicore and many-core technologies help to relieve the processing burden and enable the cloud provider to build energy-efficient and high-performance data centers. The fast developing hard-drive technologies address the challenges of massive and ever-growing data storage requirements in cloud computing. The networking technology is the fundamental backbone in cloud computing in that it connects all the cloud components together and delivers cloud services to cloud consumers. These three hardware advancements greatly empowered the development of cloud infrastructure.

Distributed computing serves as the predecessor technology that enabled the birth of cloud computing. Some essential concepts of distributed computing, such as integrating existing IT resources, are widely adopted in cloud computing. Virtualization technology abstracts the implementation of cloud platforms by encapsulating the hardware details, which makes it possible to create a dynamic number of VMs depending on the needs of the application. Virtualization plays the most important role in the implementation of cloud computing platforms, especially for the IaaS model.

The distributed file system is the core of distributed data storage technology, which is the fundamental storage model in cloud computing. The distributed storage model not only provides virtually unlimited, secured and fault-tolerant data storage and access functions, but it also enables big IT companies to build a reliable system with a group of cheap hardware. Due to these benefits, the distributed file system is playing an essential role in the development of cloud services, especially the cloud storage service.

Web-associated technologies play a critical role in the evolution of cloud computing. The Web services act as the building blocks for a cloud platform, especially in the SaaS and DaaS layers. SOA is utilized for guiding the implementation of cloud computing platforms. The Web service technology and SOA's service-oriented concept makes the on-demand service/application possible. Finally, the Web provides the ability to collaborate, integrate, present, and share information over the Internet across various communities seamlessly.

3.7 SUMMARY

This chapter introduces cloud computing enabling technologies including hardware advancements (Section 3.1), computing technologies (Section 3.2), virtualization (Section 3.3), the distributed file system (Section 3.4), and Web x.0 (Section 3.5). A conclusion discussion is provided in Section 3.6.

3.8 PROBLEMS

1. Enumerate the key enabling technologies of cloud computing from the aspects of hardware, computing, file system, and the Web.
2. Why could many-core technology help cloud providers build energy-efficient data centers?
3. Why can smart devices (e.g., smartphones and tablets) help popularize cloud services? Have you used your device(s) to access cloud services?
4. What are the major differences among cluster computing, grid computing, utility computing, and cloud computing?
5. Why is networking important to cloud computing?
6. What is virtualization and how does it help enable cloud computing? Enumerate four virtualization solutions.
7. What are the major differences between a distributed file system (DFS) and a traditional file system? Please give two examples of DFS.
8. Enumerate the generations of the Web.
9. What are the characteristics of SOA and how does SOA relate to cloud computing?
10. Think about a technology and discuss why you think it is important for cloud computing.

REFERENCES

Aghaei, S., M. A. Nematbakhsh, and H. K. Farsani. 2012. Evolution of the World Wide Web: From Web 1.0 to Web 4.0. *International Journal of Web & Semantic Technology* (IJ *WesT*) no. 3: 1–10.

Barham, P., B. Dragovic, K. Fraser, S. Hand, T. Harris, A. Ho, ... and A. Warfield. 2003. Xen and the art of virtualization. *ACM SIGOPS Operating Systems Review* 37, no. 5: 164–177.

Benslimane, D., S. Dustdar, and A. Sheth. 2008. Services mashups: The newgeneration of Web applications. *Internet Computing, IEEE* 12, no. 5: 13–15.

Borthakur, D. 2007. The Hadoop Distributed File System: Architecture and design. *Hadoop Project Web Site* no. 11: 21.

Chai, L., Q. Gao, and D. K. Panda. 2007. Understanding the impact of multicore architecture in cluster computing: A case study with Intel dual-core system. In *Cluster Computing and the Grid. CCGRID 2007. 7th IEEE International Symposium*, pp. 471–478. IEEE.

Foster, I., Y. Zhao, I. Raicu, and S. Lu. 2008. Cloud computing and grid computing 360-degree compared. In *Grid Computing Environments Workshop. GCE'08*, pp. 1–10. IEEE.

Geoffrey, R. 2009. Cloud Computing and SOA, Systems Engineering at MITRE, SOA Series. http://www.mitre.org/work/tech_papers/tech_papers_09/09_0743/09_0743.pdf (accessed April 23, 2013).

Ghemawat, S., H. Gobioff, and S. Leung. 2003. The Google File System. *ACM SIGOPS Operating Systems Review* 37, no. 5: 29–43. ACM.

Liu, H. and D. Orban. 2008 (May). Gridbatch: Cloud computing for large-scale data-intensive batch applications. In *Cluster Computing and the Grid, 2008. CCGRID'08. 8th IEEE International Symposium*, pp. 295–305. IEEE.

McEvoy, G.V. and B. Schulze 2008. Using clouds to address grid limitations. In: *Proceedings of the 6th International Workshop on Middleware for Grid Computing* no. 11. Leuven, Belgium.

Mell, P. and T. Grance. 2011. The NIST definition of cloud computing. *NIST Special Publication* 800: 145.

Pase, D. M. and M. A. Eckl. 2005. *A Comparison of Single-Core and Dual-Core Opteron Processor Performance for HPC*. IBM xSeries Performance Development and Analysis. Research Triangle Park, NC: IBM.

Sagona, P. 2009. A preliminary performance analysis of medical image registration on single-core and multicore clusters. *Lecture Notes in Computer Science*. doi 10.1.1.130.1367.

Sahoo, J., S. Mohapatra, and R. Lath. 2010. Virtualization: A survey on concepts, taxonomy and associated security issues. In *Computer and Network Technology (ICCNT), 2nd International Conference*, pp. 222–226. IEEE.

Shakian, A. 2011. Oracle Cloud File System. Oracle White Paper. http://www.oracle.com/technetwork/products/cloud-storage/cloudfs-overview-wp-279856.pdf?ssSourceSiteId=ocomen (accessed March 12, 2013).

Shvachko, K., H. Kuang, S. Radia, and R. Chansler. 2010. The Hadoop Distributed File System. In *Mass Storage Systems and Technologies (MSST), IEEE 26th Symposium*, pp. 1–10. IEEE.

Tograph, B. and Y. R. Morgens. 2008. Cloud computing. *Communications of the ACM* 51, no. 7.

Vaquero, L. M., L. Rodero-Merino, J. Caceres, and M. Lindner. 2008. A break in the clouds: Towards a cloud definition. *ACM SIGCOMM Computer Communication Review* 39, no. 1: 50–55.

Yeager, P. S. 2003. A distributed file system for distributed conferencing system. Ph.D. diss., Florida: University of Florida.

Youseff, L., M. Butrico, and D. D. Silva. 2008. Toward a unified ontology of cloud computing. *Grid Computing Environments Workshop*. GCE'08. IEEE.

Part II

Deploying applications onto cloud services

This part progressively introduces the general procedures for deploying geoscience applications onto cloud services. First, the steps for deploying a simple Web application onto cloud services are illustrated to let readers experience how to use cloud computing (Chapter 4). A database-driven application and a high performance computing (HPC) application are then demonstrated for how to cloud-enable practical geoscience applications (Chapter 5). Finally, cloud service selection strategies are introduced and discussed based on the cloud service evaluation criteria and cost models (Chapter 6).

Chapter 4

How to use cloud computing

Kai Liu, Qunying Huang, Jizhe Xia,
Zhenlong Li, and Peter Lostritto

Using a simple Web application as an example, this chapter demonstrates the basic steps for deploying applications to cloud services.

4.1 POPULAR CLOUD SERVICES

4.1.1 Introduction

Cloud services provide the computing resources (storage, network, and computation) in a virtualized fashion to consumers. The eventual success of a cloud service is determined by the applications supported and the improvements made to the applications in comparison to a traditional computing service (Dastjerdi, Garg, and Buyya 2011). Improvements can normally be achieved in application performance, cost savings, availability, stability, security, accountability, and other aspects required by specific applications (Emeakaroha et al. 2011). A well-designed process for migrating or deploying applications onto a cloud service is critical to maximize the improvements (Afgan et al. 2011).

The deployment processes are different based on the selected cloud service and the complexities of the application (Emeakaroha et al. 2011). For example, Infrastructure as a Service (IaaS) and Platform as a Service (PaaS) can support different types of applications, and the steps for deploying these applications are often different. Different applications can also have different requirements for the operating system and configuration. For example, a simple Web application may only require a Web server, but a complex application such as a financial management system may include a database, financial analyses, management tools, server side and client side software modules, and data backup components. To find out the best deployment process, consumers often start with analyzing the application requirements and available cloud services. The deployment workflow can then be designed according to the specific cloud service and application requirements (Afgan et al. 2011). This chapter demonstrates the basic steps for deploying applications onto two popular cloud services: Amazon Web Services (AWS) and Windows Azure.

4.1.2 Amazon AWS and Windows Azure

Amazon AWS and Windows Azure are becoming increasingly popular as reflected by the results of an analysis of Google Insights for Search (Figure 4.1). The number of search queries against the two terms *Amazon AWS* and *Windows Azure* has constantly increased from the end of 2008 to 2013.

As the first successful and the biggest cloud provider around the world, Amazon provides the most complete cloud services and user-friendly graphical user interfaces (GUI). It is widely adopted by many organizations to host their cloud applications. For example, Amazon became the first cloud provider to go through the Federal Information Security Management Act (FISMA) medium level certification and provides the Government Services Administration (GSA) with service for U.S. government agencies. Windows Azure has the advantage of providing both Windows operating systems and the visual studio development platform to provide PaaS.

- *AWS*—This is the name of a suite of Web services made available by Amazon Inc. It has been operating since 2006, and serves hundreds of thousands of customers worldwide. AWS is a comprehensive platform that offers a list of cloud services such as computing, storage, content delivery, database, deployment and management, networking and other services (Varia and Mathew 2012). This book uses the following AWSs: Amazon Elastic Compute Cloud (Amazon EC2, a

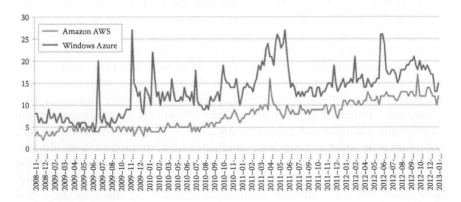

Figure 4.1 (See color insert.) Search trends among Amazon Web Services (AWS) and Windows Azure.

service providing resizable computing capacity), Elastic Block Store (EBS), and the Amazon Simple Storage Service (Amazon S3, an Internet-based storage service).

- *Windows Azure*—This is a Microsoft cloud service for the public cloud. It has four parts including Windows Azure, SQL Azure, Windows Azure AppFabric, and Windows Azure Marketplace (Chappell 2010). Windows Azure is a windows environment for running applications and storing data on computers in Microsoft data centers; SQL Azure is a cloud-based relational database service; Windows Azure AppFabric is a cloud-based infrastructure service for applications running on the cloud or on premises; and Windows Azure Marketplace is an online service for purchasing cloud-based data and applications.

Windows Azure also provides three types of execution services: Virtual Machines (VMs), Web Site Service, and Cloud Services. The ability to create a virtual machine on demand, whether from a standard image or a customized image, can be used to create a Windows or Linux VM for application deployment. Web Site Service offers a managed Web environment using Internet Information Services (IIS). Cloud consumers can move an existing IIS Web site into Windows Azure Web sites without starting a VM.

4.2 USE CASE: A SIMPLE WEB APPLICATION

A common type of application on the Internet is a Web site. Web applications are widely used for publishing geoscience applications to distributed users. HTML is the basis of a Web application. This section demonstrates deploying a Hello Cloud Web application onto Amazon AWS and Windows Azure.

4.2.1 HTML design for the Hello Cloud Web application

The Hello Cloud application is a simple file `index.html` with links to the official pages of six cloud services or solutions: Amazon AWS, Windows Azure, Apache CloudStack, Eucalyptus Cloud, Nimbus, and OpenNebula (Figure 4.2).

The HTML file content is illustrated as below.

```
<!DOCTYPE html>
<html>
<body>
<h1>Hello Cloud</h1>
```

Hello Cloud

- Amazon AWS
- Windows Azure
- Apache Cloudstack
- Eucalyptus Cloud
- Nimbus
- OpenNebula

Figure 4.2 Home page for the Hello Cloud Web site.

```
<ul>
<li><a href="http://aws.amazon.com/">Amazon AWS</a></li>
<li><a href="http://www.windowsazure.com/en-us/">Windows
    Azure</a></li>
<li><a href="http://incubator.apache.org/cloudstack/">Apache
    Cloudstack</a></li>
<li><a href="http://www.eucalyptus.com/">Eucalyptus Cloud
    </a></li>
<li><a href="http://www.nimbusproject.org/">Nimbus</a></li>
<li><a href="http://opennebula.org/">OpenNebula</a></li>
</ul>
</body>
</html>
```

4.2.2 Web servers

A Web server is needed to deliver the Web application to browsers. Three types of Web servers are used in this book: Apache HTTP server (httpd), Apache Tomcat server, and Microsoft Internet Information Services (IIS). The Apache HTTP server is one of the most popular Web servers on the Internet.[1] The Apache Tomcat server is an open-source software implementation of the Java Servlet and JavaServer Pages technologies. IIS is another popular Web server developed by Microsoft for Windows. IIS 7.5 supports HTTP, HTTPS, FTP, FTPS, SMTP, and NNTP. Section 4.3 describes how to migrate the Hello World Web application to the Apache HTTP server and Microsoft IIS running on a cloud service.

---

[1] See Apache at http://httpd.apache.org/.

4.3 DEPLOYING THE WEB APPLICATION ONTO CLOUD SERVICES

4.3.1 Amazon Web Services

Through the AWS Management Console[1] or Amazon EC2 AMI Tools,[2] cloud consumers can request to launch an instance based on a predefined Amazon Machine Image (AMI). The instance can be used as a server computer for consumers to deploy their applications.

Figure 4.3 demonstrates how to deploy the Hello Cloud application onto the Amazon EC2 platform. The details of each step are as follows:

> *Step 1. Sign up for Amazon AWS*—An AWS account is required to sign into Amazon EC2. To create an AWS account, cloud consumers can go to the AWS Web page[3] and click Sign Up, which will direct consumers to the account creation wizard. Once an AWS account is created, it can be used to access AWS (Figure 4.4).
>
> *Step 2. Authorize network access*—This step is used to: (1) enable consumers access to the instance with SSH (Secure Shell) or RDP (Remote Desktop); and (2) allow the instance to accept Web traffic on a specific port. AWS Management Console, a simple Web interface, can be used to authorize network access. Cloud consumers can get access to the console by (1) selecting the AWS Management Console after logging onto AWS (Figure 4.5), and (2) selecting EC2 under the Compute & Networking category to enter the EC2 dashboard.

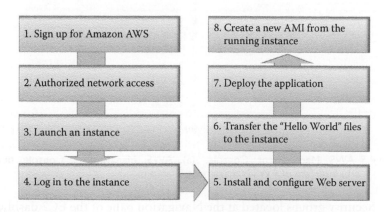

Figure 4.3 The process of deploying Hello Cloud onto Amazon EC2.

---

[1] See AWS at http://aws.amazon.com.
[2] See AWS at http://aws.amazon.com/ec2/.
[3] See AWS at http://aws.amazon.com/.

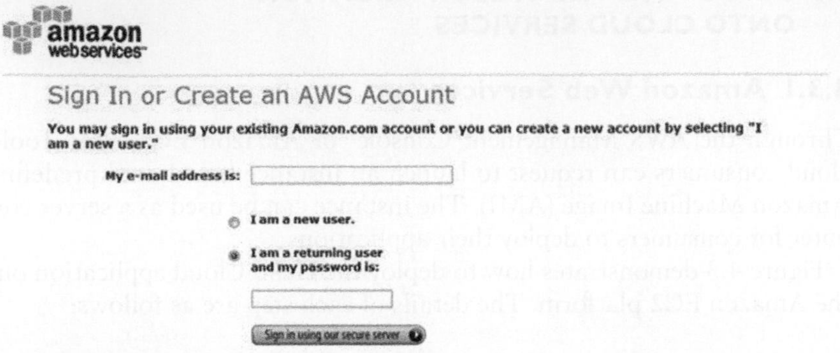

Figure 4.4 Sign in or create an AWS account.

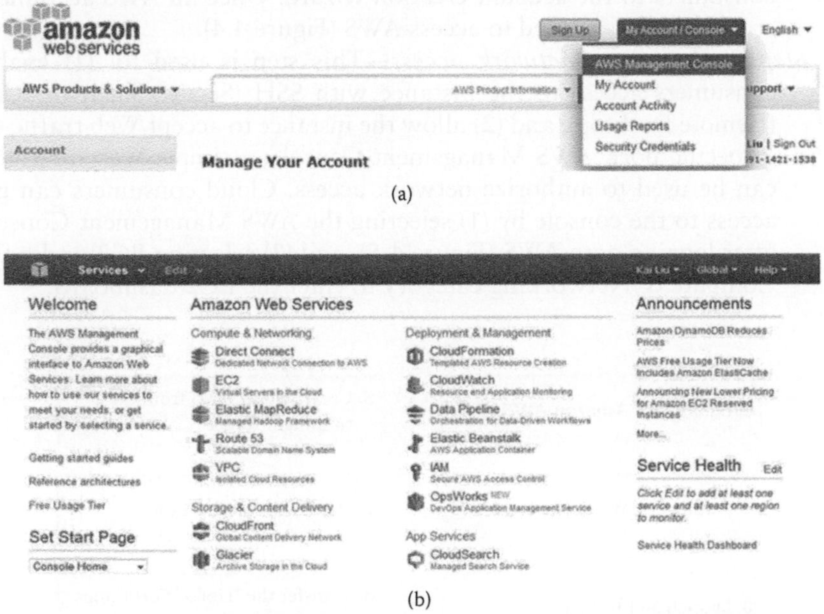

(a)

(b)

Figure 4.5 AWS Management Console. (a) AWS Management Console button; (b) Interface of AWS Management Console.

Security groups located at the Navigation pane of the EC2 dashboard (Figure 4.6) act as a firewall that controls the traffic allowed to reach the instance. Consumers can select Security Groups and then create a security group.

Figure 4.7 shows how to create a new security group to authorize the network access for the Hello Cloud application. SSH and HTTP

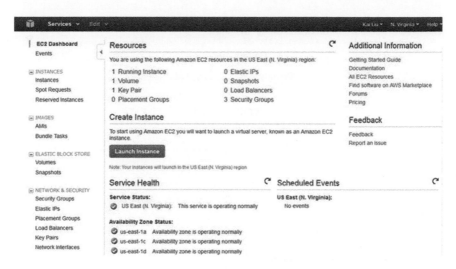

Figure 4.6 Interface of EC2 dashboard. The Security Groups menu is in the left navigation pane.

access are required for the application, so consumers need to add them accordingly. Figure 4.7b shows how to create a new rule for SSH. Consumers need to select SSH in the dropbox and input the public IP address in the CIDR (Classless Inter-Domain Routing) notation as the source. In this case, SSH from 129.174.63.89 is enabled. Figure 4.7c shows how to create a new rule for HTTP. The source 0.0.0.0/0 enables HTTP access from anywhere. Consumers need to click Apply Rule Changes after adding the rules.

Step 3. Launch an Instance—A key pair is required to launch an instance and will be used to log in to the instance after launch. Consumers need to go to the Key Pairs page in the EC2 Dashboard to create a key pair and then save the key pair on a local computer (Figure 4.8).

A public AMI with CentOS 6.3 is selected to host the Hello Cloud application. Cloud consumers can click the Launch Instance button in the EC2 dashboard (Figure 4.6) to launch an instance. After clicking the button, consumers will be directed to the wizard to select Launch an Option to launch a new instance. Consumers can directly search the available CentOS AMI with the preinstalled OS after clicking the AWS Marketplace tab and enter "CentOS" as the keyword (Figure 4.9).

Figure 4.10 shows the search results from the AWS marketplace. It lists the summaries of each result such as the AMI name, version, seller, price, and description. The first item is the official CentOS 6.3 x86_64 image. Consumers can select this AMI by clicking its title and then clicking Continue on the next page.

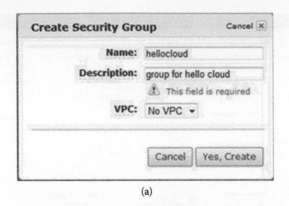

(a)

(b)

(c)

Figure 4.7 Create Security Group. (a) Create a "hellocloud" group; (b) Create a new rule for SSH; (c) Create a new rule for HTTP.

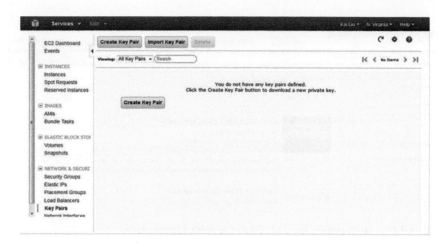

Figure 4.8 Create a key pair in the Key Pairs page.

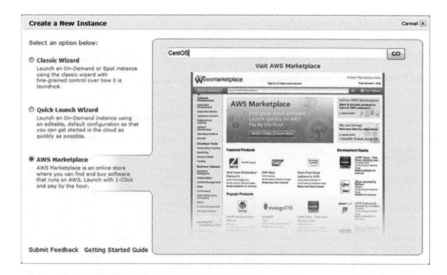

Figure 4.9 Launching a new instance by searching and using AMI from the AWS Marketplace.

After selecting the AMI, cloud consumers can select configurations for the instance (such as region, instance type, firewall settings, and key pair) and then launch the instance with one click (Figure 4.11). The security group created in Step 2 should be used as the firewall settings and the key pair created in Step 3 should be used as the key pair. If consumers prefer more options, then consumers can click Launch with the EC2 Console tab to launch Marketplace products via the EC2 Console.

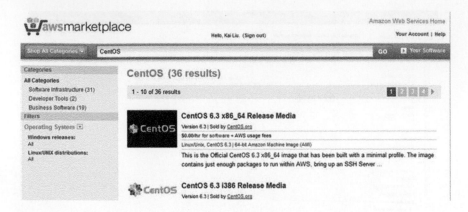

Figure 4.10 Search results for CentOS in the AWS Marketplace.

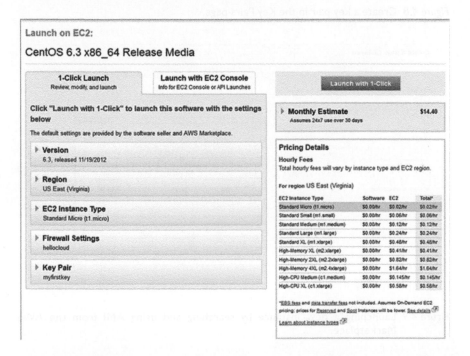

Figure 4.11 Launch the instance with one click.

Step 4. Log in to the instance—After the instance is launched, consumers can log in to it to get the full root access through the remote accessing method SSH. For Linux or Mac operating systems, users need to use the public key of the key pair, created when launching the instance, to log in to the instance in the format below.

```
$:chmod 400 ssh-keypair.pem
$:ssh -i ssh-keypair.pem username@ec2-xxx-xxx-xxx-xxx.
compute-1.amazonaws.com
```

ssh-keypair.pem is the public key file and "username" is the account name to log in to the Linux system. The default account name is "root" for Redhat Linux system, and "ubuntu" for the Ubuntu system. If the AMI is prepared by Amazon, the default account usually is "ec2-user".

PuTTY[1] can be used to connect the instance from the Windows machine. The PuTTY files can be downloaded from its official Web site.[2] In this example, PuTTYgen is needed to convert the Amazon key pair into a private key and PuTTY is needed to connect to the running instance. Before logging onto the instance, cloud consumers should load the Amazon key pair to PuTTYgen, and click Save Private Key to save the key in PuTTY's format (Figure 4.12).

With the private key converted, consumers can connect to the instance using PuTTY by entering the IP address of the instance (Figure 4.13a) in the Host name field and selecting the private key in the

Figure 4.12 Convert an Amazon key pair to a private key.

---

[1] See PuTTY at http://www.chiark.greenend.org.uk/~sgtatham/putty/.
[2] See PuTTY Download page at http://www.chiark.greenend.org.uk/~sgtatham/putty/download.html.

(a)

(b)

Figure 4.13 Log in to the instance using PuTTY. (a) Enter Host Name; (b) Select private key; (c) Type "root" as login name.

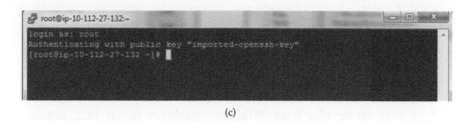

(c)

Figure 4.13 (Continued) Log in to the instance using PuTTY. (c) Type "root" as login
name.

authorization window (click Connection-SSH-Auth) (Figure 4.13b).
After clicking the Open button in PuTTY, a new shell window will be
created. Consumers need to type "root" as the username to log in to
the instance.

Step 5. Install and configure the Web server—Apache HTTP server (httpd)
is used to deploy the Hello Cloud application. The following commands
can be used to install and configure httpd on the CentOS instance.

```
$: yum install httpd
$: service httpd start
$: chkconfig httpd on
```

The first command installs httpd. The yum command is used to
automatically install the software or library on CentOS. The second
command starts the httpd server. The third command enables httpd
service to start automatically.

The instance has a default firewall, which is set by AMI providers.
Consumers need to configure the firewall accordingly. The following commands are used to configure the firewall in the Hello Cloud
application.

```
$: iptables -I INPUT -p tcp --dport 80 -j ACCEPT
$: service iptables save
```

The first command enables input Web traffic on port 80. The
second command is used to save the firewall rules.

Step 6. Transfer the Hello Cloud file onto the instance—scp[1] is used
to transfer the files from most Linux and Mac computers onto the
instance. The command can securely copy files and directories
between remote hosts. The following command can be used to transfer the Hello Cloud html file.

---

[1] See Linux at About.com at http://linux.about.com/od/commands/l/blcmdl1_scp.htm.

```
$: scp -i myfirstkey.pem index.html ubuntu@
   ec2-54-235-3-170.compute-1.amazonaws.com:/root
```

In the command, myfirstkey.pem is the private key; index.
html is the html file required to transfer; root is the name of the
account on the Amazon EC2 instance. /root is the name of the
directory on the Amazon EC2 instance.

Transfering files from Windows machines to an EC2 instance
is different. One of the easiest ways is to use WinSCP.[1] WinSCP is
an open-source widget including SFTP, SCP, FTPS, and FTP for
Windows. Its principal function is to transfer files between a local
and a remote computer. WinSCP.exe can be downloaded from its official Web site. After the installation, the host name, user name, and
private key file are needed to transfer files. The host name is the IP
address of the EC2 instance, the username is the name to log in to the
instance which was introduced in Step 4, and the private key file is a
ppk file generated by PuTTYgen.

An SFTP session will be established to connect the instance and
local Windows desktop by clicking the Login button (Figure 4.14).
Then consumers can transfer files from the local machine to the
Amazon EC2 instance (Figure 4.15).

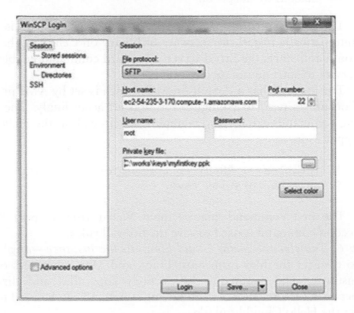

Figure 4.14 WinSCP Login.

---

[1] See WinSCP at http://winscp.net/eng/index.php.

Step 7. Deploy the application—Consumers need to deploy the Hello Cloud application under the document root directory of httpd (`/var/www`). The following commands can be used in this step.

```
$:mv /etc/httpd/conf.d/welcome.conf /etc/httpd/conf.d/
   welcomebak.conf
$:mv /root/index.html ./hellocloud /var/www/html
$:service httpd restart
```

The first line moves the default home page of httpd to another name. The second line moves the Hello Cloud home page to the document directory. The third line modifies the security contexts of the HTML folder to make it accessible from the browser. The fourth line restarts the HTTP server. The Hello Cloud application after deployment can be accessed through the EC2 URL in a browser, which looks like Figure 4.16.

Step 8. Create an AMI from the running instance—Finally, a new AMI can be created based on the instance of Hello World. If the operational instance crashes, the system can be restored from the AMI in a very fast fashion by launching an instance using the new AMI.

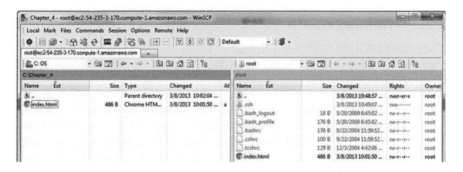

Figure 4.15 Transfer File using `WinSCP`.

Figure 4.16 Hello Cloud on Amazon Cloud.

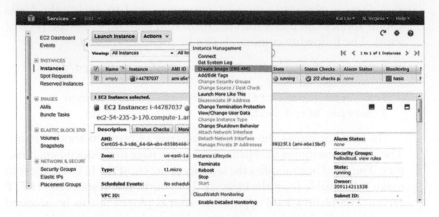

Figure 4.17 Create an AMI from a running instance.

To create an AMI from the running instance, right click the running instance in the EC2 Dashboard, and select Create Image (EBS AMI) (Figure 4.17).

4.3.2 Windows Azure

There are two ways to create a simple Web site in Windows Azure: (1) using Virtual Machine and (2) using Web Site Service. Figure 4.18 shows the steps of deploying Hello Cloud onto a Windows VM:

Step 1. Sign up for Windows Azure—In order to use Windows Azure, cloud consumers need to register an account at the Windows Azure Web site.[1] Figure 4.19 shows the Cloud Management Interface with a list of Windows Azure Services (e.g., Web sites, virtual machines, storage, database, and other services) after logging in.

Step 2. Create a virtual machine—This example uses Windows VM and IIS to deploy the simple Web application. Click the Virtual Machines tab and then click Create a Virtual Machine. Figure 4.20 shows the step to launch a Windows VM by (1) choosing the VM Image (including the system), VM size, and VM location (choose a location close to consumers to get the highest network performance) and (2) inputting the password and DNS Name. The Web site domain name will be added with the cloudapp.net. For example, if the DNS name is "hellocloudtest", the domain name for the Web site will be `http://hellocloudtest.cloudapp.net`. After clicking Create Virtual Machine, Windows Azure will create a new VM based on the inputs (Figure 4.20).

---

[1] See Windows Azure at http://www.windowsazure.com.

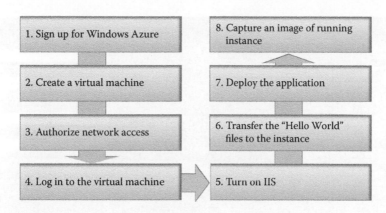

Figure 4.18 The process of deploying Hello Cloud onto Windows Azure.

Figure 4.19 Windows Azure Cloud Management Interface.

Step 3. Authorize network access—Port 80 is required by the Hello Cloud application to enable Web traffic. Cloud consumers can authorize network access by: (1) navigating to the newly created VM in the Windows Azure Preview Portal (Figure 4.22) and click the ENDPOINTS tab (Figure 4.22a) and (2) clicking the ADD ENDPOINT button at the bottom of the screen (Figure 4.22a), and opening up the TCP protocol's public port 80 as PRIVATE PORT 80 (Figure 4.22b).

Figure 4.20 Launch Windows Virtual Machine.

Figure 4.21 Windows Cloud Management Interface after Launching Virtual Machine.

Step 4. Log in to the virtual machine—After launching the VM, the virtual machine will be previewed in the Windows Azure portal. Cloud consumers can click the Connect button in the bottom of the portal (Figure 4.21) to download the RDP file for connection.

Step 5. Turn on IIS—The IIS server is used to deploy the Hello Cloud application. The IIS server can be turned on by adding roles and features in the Server Manager Dashboard after logging onto the VM (Figure 4.23).

(a)

(b)

Figure 4.22 Authorize network access. (a) ADD ENDPOINT; (b) Specify the details of the endpoint.

Figure 4.23 Turn on IIS in virtual machine.

Figure 4.24 Hello Cloud application on Windows Azure.

Step 6. Transfer Hello Cloud files onto virtual machine—Cloud consumers can simply copy and paste or drag and drop files between
local computers and VMs.

Step 7. Deploy the Application—Cloud consumers can create a folder
named "hellocloud" under the C:\inetpub\wwwroot directory in
the VM and then copy the homepage of the Hello Cloud application
to the Hellocloud folder. After the deployment, the Hello Cloud application can be accessed through the virtual machine URL in a browser
(Figure 4.24).

Step 8. Capture an image of the running virtual machine—Cloud
consumers can capture the image in the following steps: (1) Run sysprep on the VM; (2) Shut down the VM in the Cloud Management
Interface (Figure 4.21); and (3) Click the Capture button in the bottom
of the Cloud Management Interface (Figure 4.21) to capture the image.
Figure 4.25 shows the dialog box after clicking the Capture button.

4.4 CONCLUSION AND DISCUSSION

Different cloud services have different management interfaces and
functions. However, there are some general steps for using the cloud
services. Figure 4.26 illustrates the eight general steps to deploy a
simple Web application onto a cloud service based on Amazon AWS and
Windows Azure.

Step 1. Sign up for the Cloud—This is the first step for almost all public
clouds and private clouds.

Step 2. Authorize network access—Cloud consumers need to authorize
SSH or RDP access to the instance and authorize specific port(s) to
allow Web traffic.

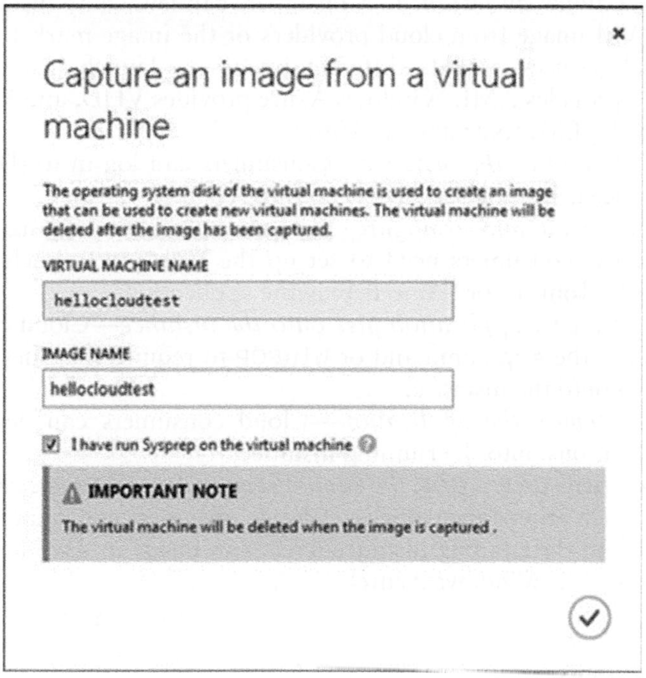

Capture an image from a virtual machine

The operating system disk of the virtual machine is used to create an image that can be used to create new virtual machines. The virtual machine will be deleted after the image has been captured.

VIRTUAL MACHINE NAME

hellocloudtest

IMAGE NAME

hellocloudtest

☑ I have run Sysprep on the virtual machine ⊘

⚠ **IMPORTANT NOTE**

The virtual machine will be deleted when the image is captured .

Figure 4.25 Capture an image of the "hellocloudtest" virtual machine. Consumers need to input the image name and check the box: "I have run Sysprep on the virtual machine."

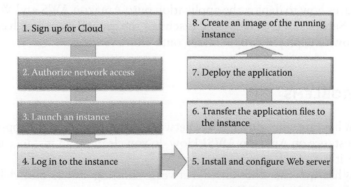

1. Sign up for Cloud	8. Create an image of the running instance
2. Authorize network access	7. Deploy the application
3. Launch an instance	6. Transfer the application files to the instance
4. Log in to the instance	5. Install and configure Web server

Figure 4.26 General steps to deploy a simple Web application onto a cloud service. The orders of the steps in the dark gray box can be changed according to different services.

Step 3. Launch an instance—Consumers can launch an instance using the VM image from cloud providers or the image market. Different cloud services provide different images to launch instances, e.g., AWS provides AMI, Windows Azure provides VHD, and Eucalyptus provides EMI (Nurmi et al. 2009).

Step 4. Log in to the instance—Consumers can log in to the running instance using SSH or RDP accordingly.

Step 5. Install and configure Web server—After logging onto the instance, consumers need to set up the Web server (such as httpd, Apache Tomcat, or IIS) to deploy the application.

Step 6. Transfer application files onto the instance—Cloud consumers can use the scp command or WinSCP to require that files be transferred onto the instance.

Step 7. Deploy the application—Cloud consumers can deploy their applications onto the running instance.

Step 8. Create an image of the running instance—An image can be used to launch an instance quickly. Cloud consumers can create an image based on the running instance (create an image in AWS and capture an image in Windows Azure).

4.5 SUMMARY

This chapter illustrates how to deploy applications onto cloud services. Section 4.1 introduces two popular cloud services: Amazon AWS and Windows Azure. Section 4.2 introduces a simple Web application. Section 4.3 introduces how to deploy the application onto Amazon AWS and Windows Azure. Section 4.4 concludes and discusses the general steps involved in the workflow of deploying applications onto cloud services.

4.6 PROBLEMS

1. What are the differences between terminating and stopping an instance on Amazon AWS? How do you clean up an instance on AWS if it is no longer needed?
2. What are the differences between shutting down and deleting a Virtual Machine (VM) on Windows Azure?
3. What are the general steps of deploying a simple Web application on Amazon EC2 and Windows Azure? What are the differences? Which one is more convenient for you?
4. How do you authorize network access in EC2 and Azure?
5. What is the role of a *key pair* in EC2?

6. What is an *instance image* in cloud computing? What is the relationship between an instance and an image?
7. How does one build an image?

REFERENCES

Afgan, E., D. Baker, A. Nekrutenko, and J. Taylor. 2011. A reference model for deploying applications in virtualized environments. *Concurrency and Computation: Practice and Experience* 24, no. 12: 1349–1361.

Chappell, D. 2010. Introducing the Windows Azure platform. http://go.microsoft.com/?linkid=9682907 (accessed August 11, 2013).

Dastjerdi, A. V., S. K. Garg, and R. Buyya. 2011. QoS-Aware deployment of network of virtual appliances across multiple clouds. In *Cloud Computing Technology and Science. (CloudCom). IEEE 3rd International Conference*, pp. 415–423.

Emeakaroha, V. C., I. Brandic, M. Maurer, and I. Breskovic. 2011. SLA-Aware application deployment and resource allocation in clouds. In *Computer Software and Applications Conference Workshops (COMPSACW): IEEE 35th Annual*, pp. 298–303.

Nurmi, D., R. Wolski, C. Grzegorczyk et al. 2009. The eucalyptus open-source cloud-computing system. In *Cluster Computing and the Grid. CCGRID'09. 9th IEEE/ACM International Symposium*, pp. 124–131.

Varia, J. and S. Mathew. 2012. Overview of Amazon Web Services. http://d36cz9buwru1tt.cloudfront.net/AWS_Overview.pdf (accessed January 23, 2013).

Chapter 5

Cloud-enabling geoscience applications

Kai Liu, Qunying Huang, and Jizhe Xia

Chapter 4 introduced the generic steps for deploying a simple Web application onto a cloud service such as Amazon EC2 and Windows Azure. However, the procedure for deploying geoscience applications onto cloud services is much more complicated. This chapter introduces the common complexities of typical geoscience applications, demonstrates how to handle these complexities when deploying geoscience applications onto cloud services, and includes two practical examples.

5.1 COMMON COMPONENTS FOR GEOSCIENCE APPLICATIONS

Deploying geoscience applications onto cloud services requires more complex hardware and software configurations than that of a simple Web application. For example, a geoscience data service may require users to log in for downloading data. This authorization process usually involves a server-side program to communicate with a database for validating user accounts. Therefore, configuring and connecting to the database are required for the deployment. Another example is geoscience simulation, which is characterized by computing intensity and data intensity. High performance computing (HPC) can be leveraged to run such simulations on multiple virtual machines (VMs) to reduce the processing time. While cloud computing excels in launching a large amount of VMs, the configuration, communication, and scheduling of these VMs should be carefully considered to achieve a decent and stable performance. In practice, server-side scripting, database, and HPC are common components that need to be considered when deploying geoscience applications onto cloud services.

5.1.1 Server-side programming

Server-side programming is a technology in which a user request is fulfilled by running a program directly to generate dynamic Web pages. A server-side program is executed on the server and the resulting Web content is passed to end users (Bradley 2013). Some popular server-side programming languages include ASP, PHP, JSP, Perl, and Ruby. Retrieving data from a database or geoscience services to generate dynamic Web content is required for many Web-based geoscience applications. Server-side programming could help developers to complete the above functions easily with the following capabilities: (1) providing secured access to geoscience services such as map layer services; (2) providing connection to interact with databases; and (3) generating new results using application logic and the latest data. Application developers can easily update a Web application by changing the information stored in the database without changing the Web pages with server-side programming. Server-side programming is required as a default component (therefore, not detailed in the following sections) for most online geoscience applications and cloud service Web consoles.

5.1.2 Database

A database is another common component in geoscience applications. Geoscience applications rely on large amounts of heterogeneous geoscience data such as Earth Observation (EO) data. A database can be used for storing, managing, and retrieving the geoscience data and metadata. Many database management systems (DBMSs) have spatial plug-ins to enable a traditional database to store geoscience data (e.g., PostGIS plug-in[1] of PostgreSQL,[2] Oracle Spatial,[3] Spatial engine of Microsoft SQL server[4]). Such kinds of databases are called *spatial databases*. Using spatial databases in geoscience applications could help reduce data redundancy, improve data access performance, and enhance data security. Databases have been widely used in many geoscience applications. For example, GEOSS Clearinghouse (Chapter 8) is based on PostgreSQL/PostGIS; Climate@Home (Chapter 9) is developed based on MySQL.[5]

5.1.3 High performance computing

Geoscience data processing and analysis is usually time-consuming, especially for large datasets. However, a fast response is required for many geoscience applications, such as dust storm forecasting and real-time routing, which needs large amounts of computational resources. A common problem for such applications is that the computational demands exceed the capacity of a traditional single processing unit. HPC provides a computing

---

[1] See PostGIS at http://postgis.net/.
[2] See PostgreSQL at http://www.postgresql.org.
[3] See Oracle at http://www.oracle.com.
[4] See Microsoft SQL Server at http://www.microsoft.com/en-us/sqlserver.
[5] See MySQL at http://www.mysql.com.

solution to this problem. Parallel computing, which is a popular way to achieve HPC, provides the capability to process larger datasets, at a higher resolution in less time (Clarke 2003). Parallel computing decomposes a serial computation task into subtasks and dispatches the subtasks onto different processors. In general, there are two task decomposition methods: (1) domain decomposition; data are partitioned and distributed across different processors, and each processor processes a subset of the data and (2) functional decomposition; each processor executes different portions of the process simultaneously. Both methods can be used for processing geoscience data, but domain decomposition is more commonly used for geoscience applications (Huang and Yang 2011; Xie et al. 2010).

An HPC system generally comprises a head node and multiple computing nodes. The computing nodes can run independently and communicate through a computer network. A middleware is installed and configured on all nodes to monitor and support communication between the head node and the computing nodes. The head node is responsible for (1) scheduling and dispatching tasks to computing nodes, (2) activating the computing tasks through configuring the middleware, and (3) collecting results from the computing nodes. Several open-source solutions can be used to deploy an HPC system (e.g., Condor,[1] MPICH2,[2] and Hadoop MapReduce[3]). Section 5.3.2 introduces how to deploy such an HPC system onto cloud services for supporting Digital Elevation Model (DEM) interpolation.

5.2 CLOUD-ENABLING GEOSCIENCE APPLICATIONS

Considering the general components discussed above, a common workflow for deploying geoscience applications onto cloud services is depicted in Figure 5.1. While this workflow is similar to the deployment of a simple Web application (Chapter 4), special considerations in the steps of "Set up environments" and "Deploy application" are highlighted.

- *Set up environments*—The VM by default contains the operating system (OS). Some software and libraries such as HTTP servers, DBMS, and Java Runtime Environments (JREs) are required by many geoscience applications. These software and libraries need to be installed and configured properly. The environment variables also need to be set up in this step. For example, *JAVA_PATH* and *PATH* are required for Java applications; *JRE_PATH* is required for the Tomcat server.

---

[1] See HTCondor at http://research.cs.wisc.edu/htcondor/.
[2] See MPICH at http://www.mpich.org/.
[3] See Hadoop MapReduce Tutorial at http://hadoop.apache.org/docs/r1.0.4/mapred_tutorial.html.

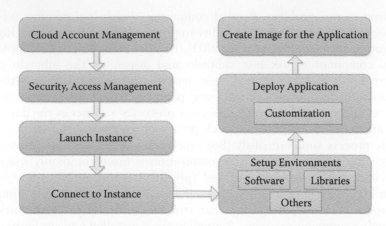

Figure 5.1 (See color insert.) General steps for cloud-enabling geoscience applications.

- *Deploy application*—The VM needs to be properly customized based on the application requirement before deploying the application, for example, customizing the database using database scripts and configuring the VM's storage service, e-mail service, and log service.

5.3 USE CASES

This section demonstrates how to deploy two typical applications onto Amazon EC2: database-driven Web applications and HPC applications.

5.3.1 Database-driven Web applications

Most cloud providers deliver the database as a service to consumers; for example, Amazon Web Services (AWS) provides Relational Database Service (RDS), DynamoDB, and SimpleDB; Windows Azure provides SQL Azure. Cloud consumers can also install their favorite DBMS on VMs instead of using database services. For example, consumers can install MySQL on a CentOS VM similar to installing MySQL on a CentOS computer.

Drupal[1] is an open-source content management framework that supports organizing, managing, and publishing Web content with a sophisticated programming interface. Geoscience applications can be built based on Drupal (e.g., the Climate@Home portal introduced in Chapter 9 and the Geoscience Platform).[2] Drupal requires a database to store the Web content, such as MySQL, PostgreSQL, and SQL server. MySQL, a popular

---

[1] See Drupal at http://drupal.org/.
[2] See Geospatial Platform at http://www.geoplatform.gov.

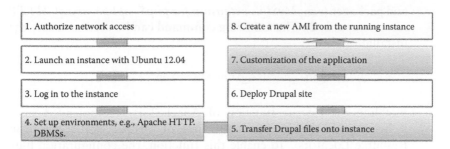

Figure 5.2 (See color insert.) The procedure for deploying the Drupal site onto EC2 (gray boxes indicate the additional steps for the deployment).

open-source database, is used in this use case. This use case demonstrates the procedure for deploying a Drupal-based Web application (backed by MySQL database) onto Amazon EC2. The steps are depicted in Figure 5.2.

Step 1. Authorize network access—Port 22 for Secure Shell (SSH) and Port 80 for HTTP should be open.

Step 2. Launch an instance—Drupal supports most Linux versions but Ubuntu is highly recommended by the Drupal community.[1] Hence, the Ubuntu 12.04 LTS image is selected to host the application. The image can be launched from the Quick Launch Wizard or the AWS Marketplace of the Amazon EC2 Management Console as detailed in Chapter 4, Section 4.3.1.

Step 3. Log in to the instance—Linux and Mac users could use the SSH command in a terminal to log in to the instance. Windows users could use PuTTY (as described in Chapter 4, Section 4.3.1).

Step 4. Set up environments—After logging onto the instance, the first step is to set up environments: installing and configuring the Apache HTTP server and the MySQL DBMS. The following commands can be used in the console.

```
$: sudo apt-get update
$: sudo apt-get install apache2
$: sudo apt-get install mysql-server mysql-client
    php5-gd
$: sudo tasksel install lamp-server
```

Here, the first line updates the system. The second line installs Apache2. The third line installs MySQL and the GD module for php5. A strong password is needed for the MySQL "root" user in the installation of MySQL. The fourth line installs related software

---

[1] See Drupal at http://drupal.org/node/850636.

and packages (e.g., *MySQL extension for php5—php5-mysql, HTTP server—apache2*). The following command can be used to secure the installation.

```
$: sudo mysql_secure_installation
```

Enabling the rewrite function of Apache2 is recommended since it helps with the correct redirection of Drupal URLs (Uniform Resource Locators). To enable this function, the configuration file/etc/Apache2/sites-available/default should be modified by replacing AllowOverride None with AllowOverride All. GNU nano[1] is recommended for modifying the file. The following command needs to be executed to activate Apache's rewrite module.

```
$: sudo a2enmod rewrite
```

After the previous steps, consumers need to (1) create a MySQL account for Drupal to connect to MySQL, and (2) create a MySQL database to store the Web application contents. The following commands create the account and database.

```
$: mysql -u root -p
mysql> create database drupal;
mysql> grant all privileges on drupal.* to ec2drupal@
  localhost identified by 'your_password';
mysql> flush privileges;
mysql> \q
```

The first line logs into the MySQL console using the root account. The second line creates a database named *drupal*. The third line grants all privileges for SQL operations (such as select, update, insert, and delete) to the user *ec2drupal*. After giving the desired privileges to the corresponding user, the flush command is called to complete the setup and to make the new settings effective. The last line exits the MySQL client.

Step 5. Transfer files onto the instance—This step transfers the Drupal installation file onto the running instance. There are two ways to transfer the file: (1) download it from the Drupal Web site[2] directly on the running instance; or (2) download the file on a local Linux/Mac/Windows machine, and then upload it to the running instance. The following command can be used to download the Drupal

---

[1] See Nano at https://help.ubuntu.com/community/Nano.
[2] See Drupal at http://drupal.org/project/drupal.

installation file from its Web site (wget[1] is supported by most Linux systems to download files from the Web).

```
$: wget http://ftp.drupal.org/files/projects/dru-
   pal-7.19.zip
```

Consumers can use the scp command or Winscp to upload files to the instance (refer to Chapter 4, Section 4.3.1). The following command can be used to upload the Drupal file.

```
$: scp -i ubuntu.pem drupal-7.19.zip ubuntu@ec2-23-
   22-98-241.compute-1.amazonaws.com:/home/ubuntu
```

Step 6. Deploy the application—After transferring the Drupal installation file onto the instance, the following commands are used to deploy the application.

```
$: sudo apt-get install unzip
$: sudo unzip drupal-7.19.zip
$: sudo mv drupal-7.19 /var/www/drupal
$: sudo chown www-data:www-data /var/www/drupal -R
$: sudo service apache2 restart
```

The first line installs the unzip utility. The second and third lines extract the Drupal files from the zip package and copy them to the default root folder of the Apache HTTP server. The fourth line changes the ownership of the Drupal files to the Apache user and group. The last line restarts the Apache HTTP server.

Step 7. Customization—Drupal's database and Web site administrators are configured through a Web interface, at http://INSTANCEIP/Drupal/install.php. INSTANCEIP is the IP allocated by AWS. Consumers can follow the wizard (Figure 5.3) to complete all the steps[2] to start the Web site.

The following example represents how to add and customize a Simple Google Maps module on this Web site. The module can be downloaded from Drupal's project Web site.[3] Once downloaded, the module's files can be extracted to the Drupal path:/var/www/Drupal/sites/all/modules. Then an administrator can log in to the Web site and enable the module on the Home-Administration-Modules page (Figure 5.4).

---

[1] See About.com-Linux at http://linux.about.com/od/comands/lblcmdl1_wget.htm.
[2] See Drupal Quick Install at http://drupal.org/documentation/install/beginners.
[3] See Drupal Simple Google Maps at http://drupal.org/project/simple_gmap.

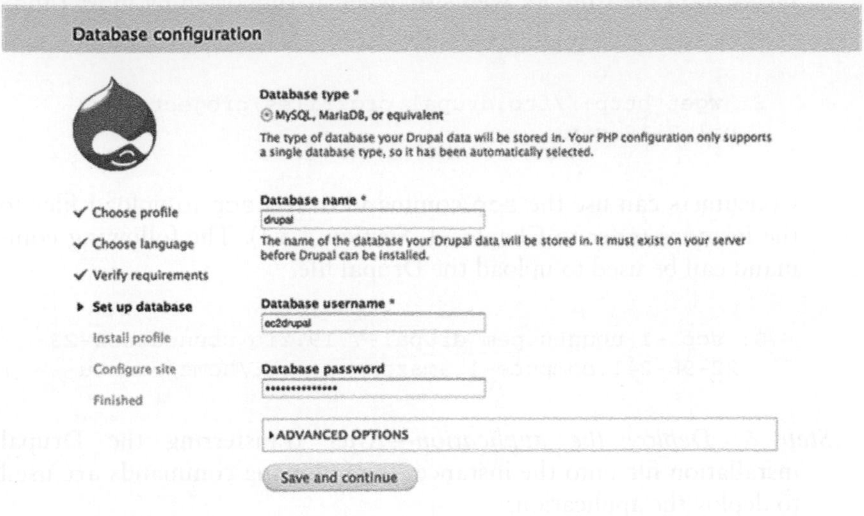

Figure 5.3 Database setting.

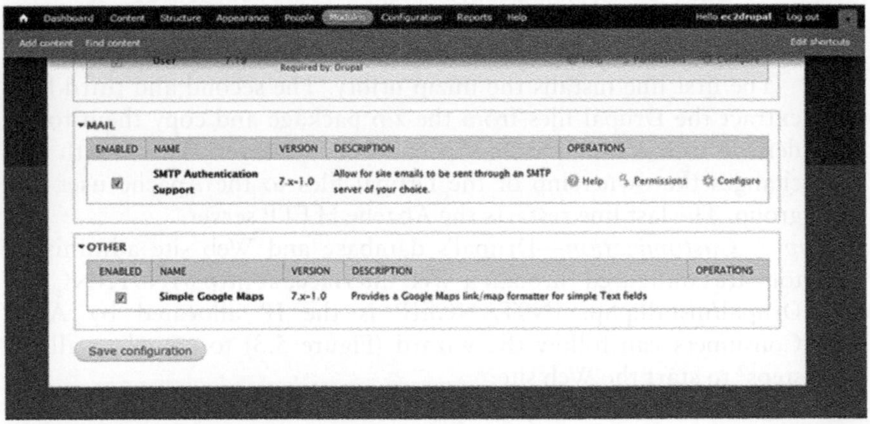

Figure 5.4 Add Simple Google Maps module.

After enabling the map module, consumers can add a field to the basic page content in the Home-Administration-Structure-Content types page. Figure 5.5 shows an example of how to add the new Google Map field.

The dispay of the new field can be managed in the Home-Administration-Structure-Content Types-Basic page (Figure 5.6).

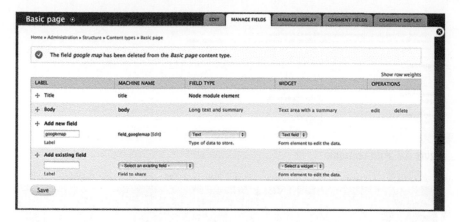

Figure 5.5 Add New Field to Article Content.

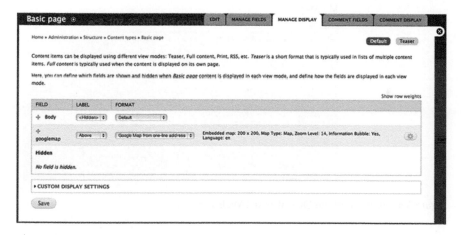

Figure 5.6 Manage Display.

The new Google Map field can be configured by clicking the "wheel" icon. Set the map size to 800*600 and the zoom level to 4.

There will be a field called *googlemap* when adding a basic page through the Home-Administration-Structure-Content Types-Basic page. The name or address (e.g., United States) should be typed in the Google Map field to setup the initial map location. To make the map appear in its own tab, check the "Provide a menu link" checkbox and enter a menu link title (Figure 5.7).

After adding Google Map, the Drupal Web site can be launched (Figure 5.8).

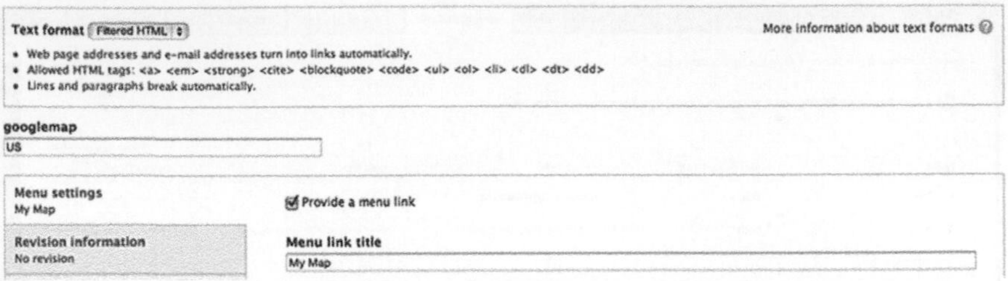

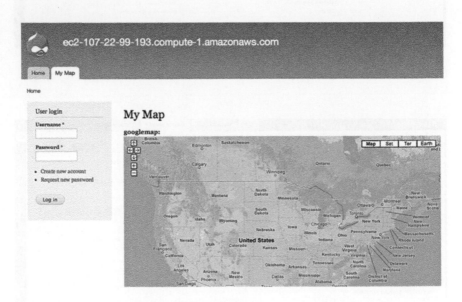

Figure 5.7 Add content.

Figure 5.8 Interface of the Drupal-based Web site.

Step 8. *Create a new AMI from the running instance*—Create this new AMI to make a full backup for this Web application (refer to Chapter 4, Section 4.3.1).

5.3.2 Typical HPC applications

This section demonstrates how to deploy an HPC application onto Amazon EC2 to support large-scale DEM interpolation. The procedures for the deployment are depicted in Figure 5.9.

Step 1. *Authorize network access* (refer to Chapter 4, Section 4.1)—The port 22 for SSH and the ports for the communication between the

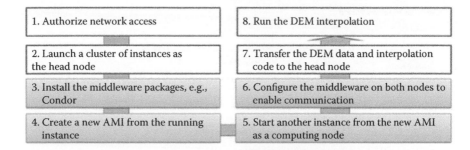

1. Authorize network access	8. Run the DEM interpolation
2. Launch a cluster of instances as the head node	7. Transfer the DEM data and interpolation code to the head node
3. Install the middleware packages, e.g., Condor	6. Configure the middleware on both nodes to enable communication
4. Create a new AMI from the running instance	5. Start another instance from the new AMI as a computing node

Figure 5.9 (See color insert.) The process for configuring an HPC system to run the DEM interpolation on EC2 (gray boxes indicate the additional steps for configuring a virtual HPC environment).

head node and the computing nodes (e.g., 9000–9999 in this case) should be opened.

Step 2. Launch an instance—EC2 provides cluster instances for running HPC applications. The *cluster instance*[1] provides relatively high CPU resources with increased network performance, making this type of instance quite suitable for HPC applications. The cluster instance can be launched from a special EBS (Elastic Block Store) backed Amazon Machine Image (AMI) using a Hardware Virtual Machine (HVM) virtualization (Figure 5.10). In addition to this type of cluster instance, users can also select High-CPU Instances or High Memory Cluster Instances depending on whether the geoscience application is data- or computing-intensive (Figure 5.11). In this case, a cluster instance with 8 CPU cores and 23 GB memory is launched (Figure 5.11).

After the head node instance is started, consumers can log in to the node through SSH. The new instance is started with an externally visible DNS name of *ec2-67-202-12-83.compute-1.amazonaws. com*, which helps to access the instance with a public key (EC2Key. pem in this case) from a local server. Please refer to Chapter 4, Section 4.3.1 for details about how to log in to an Amazon EC2 instance.

```
$ ssh -i EC2Key.pem root@ec2-67-202-12-83.compute-1.
  amazonaws.com
```

Step 3. Install the middleware packages—In this use case, Condor is the planned middleware solution, and following is the shell commands to install Condor on a CentOS 6 system.

---

[1] See AWS at http://aws.amazon.com/ec2/instance-types/.

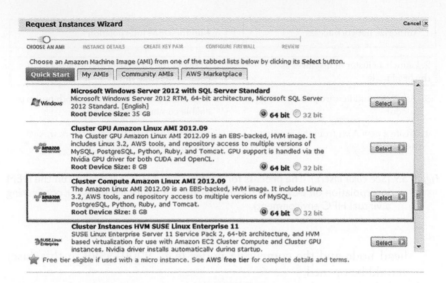

Figure 5.10 Launch cluster instances with Amazon Linux AMI.

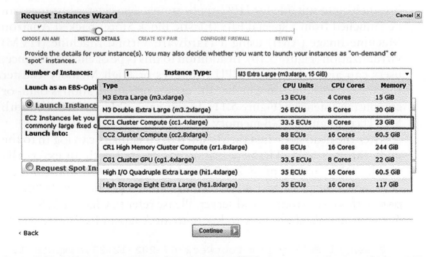

Figure 5.11 Select instance type based on the CPU, memory, and networking requirements of the geoscience HPC application.

```
$ rpm -Uvh http://download.fedoraproject.org/pub/
  epel/6/i386/epel-release-6-8.noarch.rpm ## install
  additional packages
$ yum install yum-plugin-priorities
$ rpm -Uvh http://repo.grid.iu.edu/osg-el6-
  release-latest.rpm
```

```
$ yum install condor
$ touch /etc/condor/condor_config.local ## Create
  Condor configuration file
```

After successfully installing Condor, a configuration file should be created and named *local.conf* under the directory of /etc/condor/config.d/, and the following command should be added to this file.

```
## OSG cluster configuration
# List of daemons on the node (Condor central manager
  requires collector and negotiator,
# schedd required to submit jobs, startd to run jobs)
DAEMON_LIST = MASTER, COLLECTOR, NEGOTIATOR
```

After Condor is configured, start and stop the Condor service with the following commands:

```
$ service condor start ## Start the service
$ service condor stop ## Stop the service
```

There are several important commands for using Condor (Table 5.1). The consumers can use the condor _ status to check if the head node and computing node are in the computing pool.

More information for installing Condor can be found on the Condor Web site.[1] Since the Java Development Kit (JDK) is used for compiling the DEM interpolation code, JDK must be installed on all the nodes. Instances based on Amazon Linux AMIs will have JDK preinstalled. Otherwise, the JDK (version 6, update 14) binary

Table 5.1 Condor Commands

Command	Description	Usage	Example
condor_submit	Submit a job	condor_submit [submit file]	$ condor_ submit submit.file
condor_q	Show status of jobs	condor_q [cluster]	$ condor_q 1170
condor_rm	Remove jobs from the queue	condor_rm [cluster]	$ condor_rm 1170
condor_status	Show the status of resources	condor_status	$ condor_ status-all

[1] See Condor Administration Tutorial at http://www.ccp4.ac.uk/ronan/condor_tutorials/scotland-admin-tutorial-2004-10-12.html.

package can be downloaded from Oracle at http://java.sun.com/products/archive/. After transferring the JDK package to the EC2 instance, the following commands can be used to install it.

```
[root@domU-12-31-39-13-DD-FF ~]# chmod a+x jdk-
    6u14-linux-x64.bin
[root@domU-12-31-39-13-DD-FF ~]# ./jdk-6u14-linux-
    x64.bin
```

Step 5. Create a new AMI from the running head node—Such an AMI stores the software dependencies and configurations for the HPC environment in case the head node crashes. In addition, new computing nodes can be easily added to the cluster by launching computing instances directly from this AMI.

Step 6. Launch other instances from the new AMI as computing nodes (refer to Step 2).

Step 7. Configure the middleware on all nodes to enable communication—After a computing node is started from the new AMI, further configuration on both the head node and computing nodes is needed. On the computing node, the configuration file /etc/condor/config.d/local.conf should be changed to the content below.

```
## OSG cluster configuration
# List of daemons on the node (Condor central manager
    requires collector and negotiator,
# schedd required to submit jobs, startd to run jobs)
DAEMON_LIST = MASTER, SCHEDD, STARTD
```

In addition, the computing node should have a local configuration file /etc/condor/condor _ config.local with the content shown in. The default Condor configuration file /etc/condor/condor _ config.local should be removed to avoid confusion.

```
UID_DOMAIN = $(FULL_HOSTNAME)
COLLECTOR_NAME = "OSG Cluster Condor at
    $(UID_DOMAIN)"
FILESYSTEM_DOMAIN = $(UID_DOMAIN)
ALLOW_WRITE = *.*
CONDOR_ADMIN = root@$(FULL_HOSTNAME)
CONDOR_HOST = ip-10-112-79-17.ec2.internal ## Head
    node domain name
IN_HIGHPORT = 9999
IN_LOWPORT = 9000
SEC_DAEMON_AUTHENTICATION = required
SEC_DAEMON_AUTHENTICATION_METHODS = password
SEC_CLIENT_AUTHENTICATION_METHODS = password,fs,gsi
```

```
SEC_PASSWORD_FILE = /var/lib/condor/condor_credential
ALLOW_DAEMON = condor_pool@*
NEGOTIATOR_INTERVAL = 20
TRUST_UID_DOMAIN = TRUE
START = TRUE
SUSPEND = FALSE
PREEMPT = FALSE
KILL = FALSE
```

Step 8. Transfer the DEM data and interpolation code—After the cloud-based HPC cluster is configured and launched, users can transfer the DEM data and interpolation code to the head node (refer to Section 5.3.1 for the large data transfer between the local server and the EC2 instances) and run the test. In order to submit HPC tasks to the Condor computing pool, a submission file should be created to indicate process parameters such as input directory, computing resource requirements, program and data files, and concurrent process numbers.

```
Universe = java
Executable = interpolate.class
Arguments = interpolate DEMfile.txt ## interpolate is
   the java main program, and DEMfile.txt is the input
initialdir = dir.$(Process) ## input directory
output =../interpolate.output.$(Process) ## output
   file
error = interpolate.error.$(Process)
log = ../interpolate.log
requirements = (Memory > 1024) # Select machine with
   memory size bigger than 1024Mb
transfer_input_files = MyPoint.class, PngWriter.
   class,interpolate.class, cutfile.txt
should_transfer_files = ALWAYS
when_to_transfer_output = ON_EXIT
queue 12 ## concurrent process numbers
```

The above command shows the submission file for DEM interpolation. MyPoint.class, PngWriter.class, and interpolate.class are the Java programs, and `DEMfile.txt` is the input for each task. This configuration file has 12 concurrent processes, and the command `condor_submit` (Table 5.1) can be used to submit the tasks to the cluster using a submission file named *interpolate_submit*.

```
[root@domU-12-31-39-13-DD-FF ~]# su condor # use the
   condor account
[root@domU-12-31-39-13-DD-FF ~]# condor_submit inter-
   polate_submit
```

After submitting the tasks to Condor, the command `condor _ q` (Table 5.1) can be used to check the status of all tasks. The output files `interpolate.output.X` (X indicates the process number, which is from 1 to 12 in this case) contain each task's status. Section 16.3 in Chapter 16 includes more details on the analysis of DEM interpolation result.

5.4 SUMMARY

This chapter introduces how to deploy geoscience applications onto cloud services by extending the procedure outlined in Chapter 4. Section 5.1 introduces the common requirements of a geoscience application, which includes server-side programming, database, and HPC. Section 5.2 summarizes the general steps to deploy geoscience applications onto cloud platforms. Section 5.3 demonstrates the detailed deployment process using two practical use cases. Even though different geoscience applications may rely on different technologies and, accordingly, require different hardware and software configurations, the general deployment workflow and steps introduced in this chapter are suitable for most applications.

5.5 PROBLEMS

1. Enumerate several other components required by geoscience applications besides server-side scripting, database, and high performance computing (HPC).
2. What are the general steps to deploy an application onto cloud services?
3. What is the database service provided by Amazon AWS?
4. What is the database service provided by Windows Azure?
5. How do you enable other modules such as e-mail in the Drupal case?
6. Please list five other geoscience applications.
7. How do you configure a virtual Condor HPC environment in Amazon EC2?

REFERENCES

Bradley, A. 2013. Server Side Scripting. http://php.about.com/od/programmingglossary/g/server_side.htm (accessed January 23, 2013).
Clarke, K. C. 2003. Geocomputation's future at the extremes: High performance computing and nanoclients. *Parallel Computing* 29, no. 10: 1281–1295.

Huang, Q. and C. Yang. 2011. Optimizing grid computing configuration and scheduling for geoscience analysis—An example with interpolating DEM. *Computers & Geosciences* 37, no. 2: 165–176.

Xie, J., C. Yang, B. Zhou, and Q. Huang. 2010. High performance computing for the simulation of dust storms. *Computers, Environment and Urban Systems* 34, no. 4: 278–290.

Chapter 6

How to choose cloud services: Toward a cloud computing cost model

Zhipeng Gui, Jizhe Xia, Nanyin Zhou, and Qunying Huang

There are a wide variety of cloud services, each with unique strengths and limitations. Selecting a proper cloud computing service becomes a challenge for potential cloud consumers. This chapter discusses some general criteria in selecting Infrastructure as a Service (IaaS) and Platform as a Service (PaaS) cloud services and presents a cloud advisory tool for assisting with selecting a cloud service.

6.1 THE IMPORTANCE AND CHALLENGES OF SELECTING CLOUD SERVICES

The evaluation and selection of cloud services is a critical and challenging decision-making process for cloud consumers. For example, if a consumer wants to deploy a geoscience Web portal with large amounts of geospatial data and many processing functions, the following questions may come up:

- Which cloud service should be chosen to host the application?
- What is the best virtual machine (VM) configuration that is not only capable of supporting the application but also cost-effective?
- Which cloud storage types are the best to store the geospatial data?
- Where is the best physical location to host these computing resources according to the potential users' distribution?
- How much do consumers need to pay for their applications every day, month, and year?

The complexities and challenges in cloud service selection could also result from the following perspectives:

- Different application features (e.g., data volume, data transfer speed, data communication and access frequency, and computing intensities) may have different computing resource requirements (e.g., CPU, memory, storage, network, and bandwidth).

- From an economic perspective, applications are quite different. For example, some Web applications are developed for public access, and some are just for experiments or to serve a small community. These differences prominently reflect different budget investments (fee constraints) and expectations on cloud services.
- From a cloud service perspective, commercial and open-source cloud services adopt different IT technologies (e.g., virtualization and storage) and have their own unique strengths and weaknesses in computational capacities.
- Meanwhile, various pricing models (e.g., on demand/reserved/bidding mode) complicate the choices.

The interweaving of these factors makes cloud service selection extremely time consuming and challenging. It is not only a technical problem but also a management challenge, which involves the trade-off between business expectations, investment, and capacity provisioning and application requirements. Therefore, making an optimized choice is not easy for the novices who do not have prior cloud computing experience or knowledge, nor even for experienced cloud consumers.

6.2 THE FACTORS IMPACTING CLOUD SERVICE SELECTION

To make wise decisions, cloud consumers should understand cloud measurement criteria. Various measurement criteria have been proposed (Haddad 2011; Kulkarni 2012; Repschläger et al. 2011; Rodrigues 2012). This section categorizes and introduces these basic factors in three categories including cloud service capacity provisioning, pricing rules, and application requirements.

6.2.1 Cloud service capacity provisioning and measurements

Capacity provisioning is an important factor to evaluate cloud services and the criteria include (CSMIC 2011; Rodrigues 2012):

- *Computational capability* is an essential measurement for cloud services, and contains: (1) computing capability (e.g., CPU/GPU core number and speed, memory size, and VM number); (2) storage capability (e.g., volume, I/O speed, durability, and types); and (3) network capability (bandwidth and network types).
- *IT security and privacy* (Jansen and Grance 2011) are critical concerns for cloud consumers. The Federal Information Security

Management Act (FISMA) as well as Federal Risk and Authorization Management Program (FedRAMP) have been developed as a policy guidance to address information security issues. Therefore, the capabilities to protect and guarantee security and privacy can be measured by checking if the cloud providers have complied with security-related certifications, such as PCI or SAS 70, industrial and governmental regulations and laws, and if they have been audited by the third parties.

- *Reliability and trustworthiness* (Badger et al. 2012)—Reliability describes the certainty that the consumer can be served with available cloud services. It is important to know what Service Level Agreements (SLAs) are available and how these committed agreements are kept. In addition, the trustworthiness describes the provider's infrastructural features, which may be evidence of high reliability. These include services for disaster recovery and redundant sites.

- *Customization degree and flexibility/scalability* (CSMIC 2011)—The on-demand self-service and rapid elasticity features rely on flexibility and customization capability. Customization degree refers to the capability to customize the configuration of cloud services to meet the cloud consumers' requirements. It can be measured by the available number of configuration types (e.g., VM types). Some providers even offer fully customizable VMs. For example, consumers can specify the CPU, memory, local disk, and other hardware configurations of a VM within reasonable value ranges. In addition to the hardware configuration, the software configuration, such as the supported Operating System (OS) (e.g., types and versions), are also important measurements. As the demands on computing resources are mutable for a geospatial application, the capability to dynamically change the computing capacity provisioning (i.e., scalability) is critical. Two types of scalability are usually considered: scale up and scale out. Scale up refers to the capability to upgrade individual VM instances by adding more memory, extra CPUs, or more storage space. Scale out refers to quickly provisioning new VM instances for intensive tasks and concurrent requests.

- *Manageability, usability, and customer service* (CSMIC 2011)—Governance functions for coordinating and monitoring cloud services (e.g., individualization and interactivity of the Web interface) are important features. A good cloud service should be user-friendly, that is, easy to navigate, convenient to use, and be accompanied by comprehensive monitoring functions (e.g., performance, usage, and cost). Customer service should include technical assistance via telephone, e-mail, online live chat, and other methods (knowledge base and user forums). Good customer service could help consumers learn how to utilize and manage cloud services. Free trials are another important measurement that can help consumers try and test the services themselves before deciding to purchase the service.

- *Geolocations of cloud infrastructures* not only impact the computing accessibility and performance, but also reflect some policy constraints. Both cloud consumers and users may expect the cloud resources to be closer to their location for better performance and controls. In addition, a data provider may have to comply with international, federal, or state regulations that prohibit the storage of data outside certain physical borders (Badger et al. 2012).

6.2.2 Cloud platform pricing rules

Choosing a cost-effective cloud service that fits into a cloud consumer's needs requires a cost model to consider:

- *Cost for employing VMs*—The cost is determined by the VM number, VM types, payment types, intensity of utilization, and tenancy time. Cloud providers usually offer multiple predefined VM types or fully customizable VMs that allow consumers to configure according to their application requirements. These VM types are different in both hardware and software configurations, so they offer different computing capacities and generate different tenancy costs.

There are several widely used payment types: (1) The so-called *pay-as-you-go mode* lets consumers pay for compute capacity on demand by the hour rather than by long-term commitments. (2) *Reserved mode* (*Prepaid* by Windows Azure and *Subscription* by CloudSigma) gives the option to make a low, one-time payment to reserve cloud resources for fixed time periods and in turn receive a significant discount against the hourly charge mode. (3) *Purchasing units* is a unit-based billing mode (e.g., FlexiScale[1] and OpSource[2]). In this mode, cloud consumers buy credit units and use the units on the cloud resources consumed (e.g., VM, storage, network, and software images). For example, FlexiScale charges 16 units per hour for a VM with 4 CPU cores and 6 GB RAM size. And (4), the *biding mode* enables consumers to bid for unused capacity (e.g., Amazon EC2's Spot Instances).

The intensity of utilization on both a VM and network is also a cost factor. For example, Amazon classifies *Light*, *Medium*, and *Heavy* intensity levels for reserved instance types. Different intensities will be charged with different unit prices.

---

[1] See FlexiScale at http://www.flexiscale.com/.
[2] See OpSource at http://www.opsource.net/.

- *Cost on data transfer*—The price of data transfer is based on data transferred *in* and *out*, geolocation (region/zone partition), and also data volume. An Internet data transfer in/out fee is charged when data are transferred inbound/outbound of the cloud service or across regions. Regional data transfer fees are charged when data are transferred across zone boundaries but within the same cloud provider. For example, data transferred between AWS instances located in different availability zones within the same region will be charged with a regional data transfer. Data transferred between AWS instances in different regions will be charged as an Internet data transfer on both sides of the transfer.

- *Cost on data storage/database*—Data storage/database fees are determined by data size, storing time period, and storage types. Vendors may provide multiple storage types according to different application requirements and purposes. For example, Amazon Elastic Block Store (EBS) is designed specifically for Amazon EC2 instances in the same availability zone. Amazon Simple Storage Service (S3) is designed to store and retrieve any amount of data, at any time, and from anywhere on the Web. S3 offers Standard Storage, Reduced Redundancy Storage (RRS), and Glacier storage. Standard Storage provides a highly reliable storage infrastructure designed for mission-critical and primary data storage (99.999999999% durability and 99.99% availability); RRS offers a lower reliability (99.99%) than Standard Storage and is for storing reproducible data (e.g., thumbnails, transcoded media, or other processed data). Glacier storage is for infrequently accessed data (e.g., digital media archives and long-term database backups).

- *Other costs*—Internet Protocol (IP) address fees are charged for extra IP address and domain names associated with the VM instances. Extra network fees include the cost for virtual networks, content delivery networks (CDN), and others. Communication fees refer to the extra fees to access the cloud infrastructures (e.g., Amazon charges the requests sent to S3). In some cloud services, customer service also requires a support fee.

6.2.3 Application features and requirements

Application features and requirements dominate the selection of cloud services and sometimes influence the cloud service capacity provided. It should be carefully considered in cloud service selection. Multiple Amazon cloud storage types (described in Section 6.2.2) are good examples of how usage purposes (storing data archives or application raw/result data) and capacity demands (on security, reliability, I/O speed, and access frequency) impact the provider capacity provisioning.

The intensity feature of the application is another example. *Data intensive applications* require a large storage pool to manage and process Big Data. Data volume, I/O performance, data indexing, data backup, security, and reliability are critical capabilities. *Computing intensive applications* require powerful computing capacities (e.g., high CPU speed, more CPU cores, and large memory size) to execute large-scale computing. The application may utilize the computing power of multiple machines to improve the performance; therefore, the network configurations and communication optimizations are critical. The Amazon EC2 HPC instance type offers an option to better support this category of applications. *Concurrent intensive applications* normally involve intensive concurrent requests/responses from users or existing Web services. A large bandwidth is critical to provide decent communication performance through the Internet. Load balancing is another capability to manage concurrency intensity, which leverages multiple and sometimes distributed instances of applications in a collaborative fashion to balance end user requests and provides a single Web access point. *Communication intensive applications* (or so-called borderless applications) are highly dependent on third-party services that are outside the organizational and geographic "borders" of the applications. Just like concurrent intensive applications, a large bandwidth is required for rapid network communication, and load balancing is required for robust reliability.

6.3 SELECTING CLOUD SERVICES USING THE EARTH SCIENCE INFORMATION PARTNERS (ESIP) CLOUD ADOPTION ADVISORY TOOL AS AN EXAMPLE

It is highly desirable to design and develop a Web-based advisory tool to assist consumers from the geoscience domain to compare and select the most suitable cloud service. Such a tool should integrate knowledge about cloud services and be capable of recommending a cloud service to achieve both cost-efficiency and high performance. An ideal Web-based advisory tool should provide the following capabilities: (1) Assist cloud consumers to easily select the best solutions based on their application requirements; (2) Help cloud novices understand the basic concepts of cloud computing services, technologies, providers, and potential applications; and (3) Automatically collect and manage the pricing and configuration information of multiple cloud vendors. To verify the feasibility of advisory tools for cloud computing adoption, we designed and developed a Web-based cloud computing adoption advisory tool[1] for the Federation of Earth Science Information Partners (ESIP).

---

[1] See Cloud Computing Advisory Adoption Tool at http://swp.gmu.edu:8080/ESIPCostModelProject.

6.3.1 Architecture of the advisory tool

The architecture of the advisory tool (Figure 6.1) consists of four major components including the solution creator, solution evaluator, cloud information database, and Web GUI.

- *Solution creator* generates feasible solutions and returns recommended solutions.
- *Solution evaluator* calculates the potential tenancy cost and assesses the suitability of generated solutions.
- *Cloud information databases* store the collected cloud service information, including pricing rules, configuration scheme, and capability declarations.
- *Web GUI* controls the user interaction, solution presentation, and visualization. Cloud consumers specify their application requirements and constraints through the application requirement description wizard. Solution tables and charts are used to display and compare cloud services.

6.3.2 The general workflow for cloud service selection

For a cloud consumer, the general procedure for selecting a cloud service includes the following four steps: determining the application type, defining the application requirements and features, searching matched cloud services, and comparing services. The following paragraphs use the ESIP Advisory Tool as an example to describe the workflow in detail for cloud service selection.

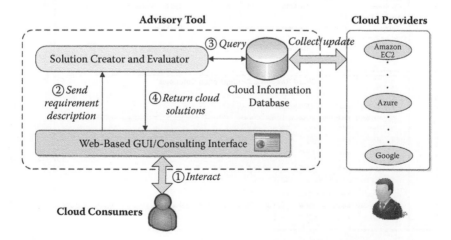

Figure 6.1 Architecture and interaction workflow of the advisory tool.

The first important step for cloud service selection is to understand the type of application to be deployed on the cloud. Usually, the type of application determines the computational features for the hardware and software configuration of the cloud services. These common geospatial application types include Web-based application, scientific computing, data processing application, and others. In the advisory tool, three application types are predefined as templates for consumers (Figure 6.2): (1) *Data Storage Application* is for storing data on cloud storage, where no application will be deployed. (2) *Simple Web Application* is suitable for small/medium-scale Web applications, such as geospatial Web portals and services (e.g., Web map services). (3) *Simple Computing Application* is for computing intensive requirements (e.g., dust storm forecasting). For experts who have sufficient knowledge on cloud computing and the application parameters, the advisory tool also provides a *Customized Application* as a template.

The following three parts must be carefully considered for application requirements and features including hardware requirements, application features, and purchase preferences (Figure 6.3). Hardware requirements can either be the basic or best hardware configurations to run the application (e.g., CPU core number, CPU speed, memory size, and local disk size). Application features describe the computational features and

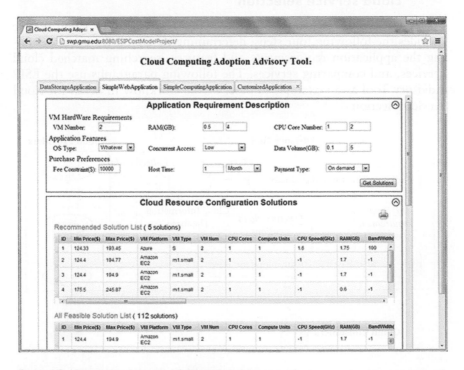

Figure 6.2 Main GUI of the cloud selection advisory tool.

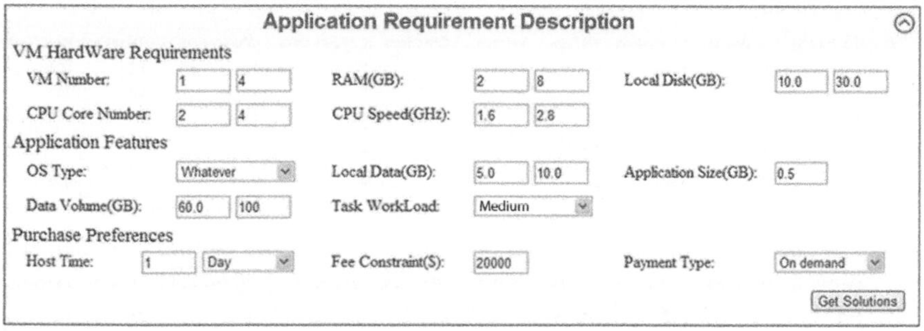

Figure 6.3 Application requirement description for simple computing application.

software requirements of the applications (e.g., OS types and version, user concurrence intensity, application and data size, network traffic and speed, and data durability requirement). Purchase preferences refer to how consumers prefer to pay (e.g., payment type, tenancy time, and cost constraints). In the ESIP advisory tool, through the predefined application type templates, consumers can directly specify application requirements and features on the Web GUI of the advisory tool. Through the template of *Customized Application*, a consumer can specify the most comprehensive requirements and feature descriptions, and the advisory tool will also provide an accurate evaluation and cost estimation.

After defining the application requirements and features, the consumer can use them as constraints to search and get feasible cloud service solutions. This process is time consuming. The consumer must collect all the configurations, pricing, and technical information for all cloud services, and then filter them using specified application requirements and features, and generate all feasible cloud service solutions.

To make a wise selection from all the possible solutions, the consumer should conduct some further analyses on cost, configuration, computing capacity, provider reputation, and others. Therefore, how to intuitively compare them is significant. To facilitate cloud solution selection, the advisory tool dynamically generates cloud solutions according to user inputs and provides an interface for visually comparing the solutions on cost, configuration, and VM computing capacity. When cloud solutions are found, two tables and three charts will be presented on the Web GUI. (1) The table Recommended Solution List lists the recommended solutions, while the All Feasible Solution List lists filtered feasible solutions (Figure 6.4). The configuration and cost of the cloud solutions are presented as table columns. Users can sort the solutions list by any of the table columns. This will help cloud consumers rank solutions by different criteria. (2) The Minimum Fee Chart (Figure 6.5 top) and Maximum Fee Chart (Figure 6.5 bottom) give consumers clear information to compare the potential fee ranges of

Recommended Solution List (5 solutions)

ID	Min Price($)	Max Price($)	VM Platform	VM Type	VM Num	CPU Cores	Compute Units	CPU Speed(GHz)	RAM(GB)	BandWidth(
1	29.04	50.5	Azure	M	4	2	2	1.6	3.5	200
2	28.63	74.08	Azure	M	4	2	2	1.6	3.5	200
3	30.13	76.58	Azure	M	4	2	2	1.6	3.5	200
4	43.8	65.26	Amazon EC2	m1.large	4	2	2	-1	7.5	-1
5	44.4	65.86	Azure	L	4	4	4	1.6	7	400

All Feasible Solution List (51 solutions)

ID	Min Price($)	Max Price($)	VM Platform	VM Type	VM Num	CPU Cores	Compute Units	CPU Speed(GHz)	RAM(GB)	BandWidth(
1	45.3	67.76	Amazon EC2	m1.large	4	2	2	-1	7.5	-1
2	43.8	65.26	Amazon EC2	m1.large	4	2	2	-1	7.5	-1
3	44.71	91.67	Amazon EC2	m1.large	4	2	2	-1	7.5	-1
4	76.02	98.48	Amazon EC2	m1.xlarge	4	4	8	-1	15	-1
5	74.52	95.98	Amazon EC2	m1.xlarge	4	4	8	-1	15	-1

Figure 6.4 Recommended Solution List and All Feasible Solution List.

the recommended solutions. By dividing the fee cost (shown as columns) of a certain solution into multiple parts (VM fee, storage fee, and data transfer fee), the adopters can intuitively understand the percentages they pay for each fee part. (3) The Virtual Machine Configuration Comparison Chart (Figure 6.6) shows the VM configuration (CPU core number, virtual compute unit number, CPU Speed, RAM size, bandwidth, local disk size) of the selected solutions as a line series chart. It helps consumers compare VM capacity parameters.

6.3.3 Use case

We use a computing application example to show how to use the advisory tool to find cloud solutions. The interaction workflow includes four steps:

> *Step 1. Application Requirement Descriptions*—A potential consumer specifies the *Application Requirement Descriptions* for the potential computing application in the *Simple Computing Application* page. Since the computing application requires more powerful computing capacity than a simple Web application, the application requirement parameters are specified (Figure 6.3). For example, the CPU core count is set to be at least 2 and preferably 4; the intensity of the computing task is medium; the tenancy time is one day; and the payment type is on-demand.
>
> *Step 2. Get Solutions*—The potential consumer clicks the Get Solutions button to generate solutions.

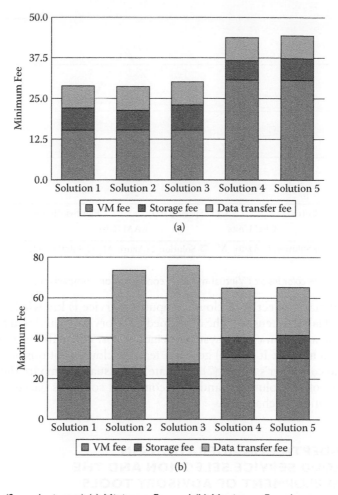

Figure 6.5 (See color insert.) (a) Minimum Fee and (b) Maximum Fee charts.

Step 3. Tables and Charts—The potential consumer interacts with the solutions in the tables and charts. Five recommended solutions and more feasible solutions are listed in the two table components (Figure 6.4). The minimum and maximum fee charts (Figure 6.5) illustrate that the recommended solutions of No. 4 and No. 5 have a larger cost proportion on VM tenancy. The VM configuration comparison chart (Figure 6.6) illustrates that the solution of No. 5 (the polyline with pink color [see Color Insert] in the line series chart) offers the most powerful computing capacity.

Step 4. Print—The consumer outputs results into a PDF file by clicking the Print button.

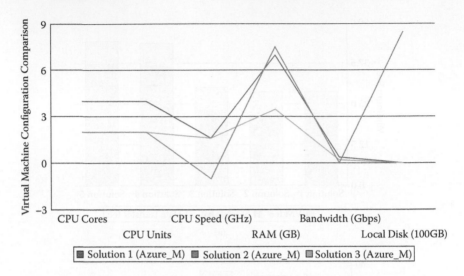

Figure 6.6 (See color insert.) Virtual machine configuration comparison.

In summary, selecting a cloud computing service is knowledge-, experience- and labor-intensive. The developed advisory tool validated the design of a cloud adoption expert system including cost calculations and suitable evaluation models. It can (1) generate feasible cloud solutions according to the cloud consumer's input; (2) calculate and visually compare the cost and capacities of the solutions, and (3) recommend solutions based on embedded evaluation mechanisms.

6.4 IN-DEPTH CONSIDERATIONS IN CLOUD SERVICE SELECTION AND THE DEVELOPMENT OF ADVISORY TOOLS

Much research has been conducted on selecting cloud services (Badger et al. 2012). Specifically, cloud service metrics[1] (CSMIC 2011; NIST 2012) start as an initiative to establish a consistent and operable measurement to enable the stakeholders of cloud services to communicate more efficiently. Cloud evaluation and selection models (Andrzejak, Kondo, and Yi 2010; Calheiros et al. 2011; Repschläger et al. 2011; Stantchev 2009) provide theoretical methods to assist decision making. Monitoring systems[2] and selection assistant systems/Web sites[3] are developed (Goscinski and Brock 2010; Martens, Teuteberg,

---

[1] See NIST RATAX Cloud Metrics Sub Group at http://collaborate.nist.gov/twiki-cloud-computing/bin/view/CloudComputing/RATax_CloudMetrics.
[2] See Global Provider View, CloudSleuth at https://cloudsleuth.net/global-provider-view.
[3] See FindTheBest, Cloud Computing Providers at http://cloud-computing.findthebest.com.

and Gräuler 2011) to maintain real-time cloud service information. Although great progress has been made on the aspects previously discussed, there are still challenges remaining which need to be carefully addressed.

6.4.1 Correctness and accuracy of evaluation models

Without high-quality evaluation models, cloud solutions are not very accurate. Having said that, establishing reliable models and validating them is a big challenge.

- *Cost calculation model*—Although variations exist on the calculation formula according to different providers and applications, a general model for basic applications should consider all charging parts described in Section 6.2.2. Meanwhile, according to the uncertainty and mutability on practical cloud usages, it is impossible to make a precise prediction on the total cost. However, the cost calculation model can provide cost details for each part to help cloud consumers clearly understand the composition of the potential cost (Li et al. 2009) and the return on investment (ROI).[1] Furthermore, a good calculation model should also have the capability of making reasonable estimations on uncertainties based upon the application types, features, and potential usage.
- *Suitable evaluation models and basic selection principles*—To make a wise selection, multiple factors (e.g., platform capacity provisioning, cost constraints, application requirements, and features) must be carefully considered. The evaluation model should be able to adjust itself to best fit the features and requirements of different application types. The basic principles to be considered in the evaluation and selection are listed as the following:
- Satisfy all constraints from the cloud consumer (e.g., cost, OS types, VM configurations, geolocation).
- Minimize the fee cost.
- Maximize the computation capacity provisioning.
- Comply with the consumer's preferences (e.g., specified vendors and tenancy types).
- Consider the spatiotemporal impacts (e.g., potential geolocation distribution of cloud resources and application users, access frequency, and time distribution).

With the in-depth consideration on spatiotemporal features, further research can be conducted to optimize the spatiotemporal distributions of

---

[1] See Microsoft Windows Azure Platform TCO Calculator at http://www.microsoft.com/brasil/windowsazure/tco/.

cloud resources (Yang et al. 2011) the for consumers to improve performance and reduce the tenancy cost (e.g., trying to avoid unnecessary data transfer fees). For example, Andrzejak et al. (2010) proposed a probabilistic model to help consumers bid optimally on Amazon Spot Instances to reach different objectives with desired levels of confidence.

6.4.2 Up-to-date information of cloud services

The pricing rules of cloud providers are frequently updated based on the usage and business strategies. Keeping the pricing up-to-date is crucial to cloud service selection systems. Without the updated information, the recommended solutions and their configurations are unreliable. Therefore, advisory systems should have the capability of updating the information in near real-time mode and in an automated process once cloud providers update offers. One potential solution is to utilize the notification functions provided by the cloud providers to obtain the information update events and trigger the crawling and updating process. Other solutions include feedback mechanisms to allow consumers and providers to update information.

6.4.3 Interactivity and visualization
functions of the advisory tool

A well-designed and user-friendly advisory system can help cloud consumers reduce the time and effort spent on information collection, and make wise selections through human–computer interaction. Currently, most systems and Web sites are still in their early stages. Cloud information collection and presentation are their primary functions.

To present information intuitively, and eventually to help consumers make wise selections, more advanced interaction and visualization functions should be utilized, especially given the popularity of visual analytics. A new way could be developed to conduct information/data visualization and human–computer interaction to solve problems, by enabling synergistic human and computer interactions. Besides, the system should also satisfy the requirements of different consumer levels (experienced and inexperienced) and help them get their results with a pleasant learning process.

6.5 SUMMARY

This chapter introduces cloud service metrics and important criteria for cloud service evaluation and selection. Section 6.1 introduces the importance of cloud service selection. Section 6.2 introduces the criteria for selecting cloud services. Section 6.3 demonstrates how to select a cloud service using the ESIP cloud advisory tool as an example. Section 6.4 discusses in-depth considerations and research on (1) the cost model and cloud evaluation, (2) up-to-date information and related collecting and updating methods, and (3) the functional design of a cloud advisory tool.

6.6 PROBLEMS

1. What are the challenges for cloud consumers in selecting a cloud service?
2. What are the major factors that impact cloud service selection?
3. Why are the features and deployment requirements of an application important to cloud service selection?
4. For the IaaS and PaaS cloud services, what kinds of services (e.g., virtual machine, storage, data backup, data transfer) are charged? Which services have a relatively stable amount of fees in certain periods of time and which ones are more mutable?
5. Please enumerate the measurements for cloud service capacity provisioning.
6. What factors usually impact the total cost when employing VMs?
7. What are the basic principles for selecting a cloud service?
8. Are cloud storage types and cloud infrastructure locations important to cloud service selection? Why?

REFERENCES

Andrzejak, A., D. Kondo, and S. Yi. 2010. Decision model for cloud computing under SLA constraints. *Modeling, Analysis & Simulation of Computer and Telecommunication Systems (MASCOTS), IEEE International Symposium* http://hal.archives-ouvertes.fr/docs/00/47/48/49/PDF/ec2pricesA_final.pdf (accessed March 29, 2013).

Badger, L., T. Grance, R. Patt-Corner, and J. Voas. 2012. Cloud computing synopsis and recommendations. *NIST Special Publication* 800: 146.

Calheiros, R. N., R. Ranjan, A. Beloglazov, C. A. F. De Rose, and R. Buyya. 2011. CloudSim: A toolkit for modeling and simulation of cloud computing environments and evaluation of resource provisioning algorithms. *Software: Practice and Experience* 41, no. 1: 23–50.

CSMIC. 2011. Service Measurement Index (Version 1.0). Silicon Valley, Moffett Field, CA: Carnegie Mellon University. http://csmic.org/wp-content/uploads/2011/09/SMI-Overview-110913_v1F1.pdf (accessed March 12, 2013).

Goscinski, A. and M. Brock. 2010. Toward dynamic and attribute based publication, discovery and selection for cloud computing. *Future Generation Computer Systems* 26, no. 7: 947–970.

Haddad, C. 2011. Selecting a Cloud Platform: A Platform as a Service Scorecard. WSO2 White Paper. http://wso2.org/library/whitepapers/2011/12/selecting-cloud-platform-platform-service-scorecard (accessed March 29, 2013).

Jansen, W. and T. Grance. 2011. Guidelines on security and privacy in public cloud computing. *NIST Special Publication* 800: 144.

Kulkarni, P. 2012. Guidelines for Selecting Cloud Provider and Determining Cloud Type. http://blog.harbinger-systems.com/2012/02/guidelines-for-selecting-cloud-provider-and-determining-cloud-type (accessed March 12, 2013).

Li, X., Y. Li, T. Liu, J. Qiu, and F. Wang. 2009. The method and tool of cost analysis for cloud computing. In *Cloud Computing. CLOUD'09. IEEE International Conference*, pp. 93–100.

Martens, B., F. Teuteberg, and M. Gräuler. 2011. Design and implementation of a community platform for the evaluation and selection of cloud computing services: A market analysis. *ECIS Proceedings*, 215.

NIST. 2012. NIST Cloud Computing Reference Architecture Cloud Service Metrics Description (Draft). http://collaborate.nist.gov/twiki-cloud-computing/pub/CloudComputing/RATax_CloudMetrics/RATAX-CloudServiceMetricsDescription-DRAFT-v1.1.pdf (accessed March 12, 2013).

Repschläger, J., S. Wind, R. Zarnekow, and K. Turowski. 2011. Developing a cloud provider selection model. In *Proceedings of Enterprise Modelling and Information Systems Architectures: Proceedings of the 4th International Workshop on Enterprise Modelling and Information Systems Architectures. EMISA*. Hamburg, Germany, September 22–23. (GI publishes recent findings in informatics [i.e., computer science and information systems] as a series to document conferences that are organized in co-operation with GI and to publish the annual GI Award dissertation), p. 163.

Rodrigues, T. 2012. Comparing Cloud Infrastructure-as-a-Service Providers (11 cloud IaaS providers compared). http://www.techrepublic.com/blog/datacenter/11-cloud-iaas-providers-compared/5285 (accessed March 12, 2013).

Stantchev, V. 2009. Performance evaluation of cloud computing offerings. In *Advanced Engineering Computing and Applications in Sciences. ADVCOMP'09. 3rd International Conference*, pp. 187–192.

Yang C., M. Goodchild, Q. Huang, D. Nebert, R. Raskin, M. Bambacus, Y. Xu, and D. Fay. 2011. Spatial cloud computing—How can geospatial sciences use and help to shape cloud computing? *International Journal of Digital Earth* 4, no. 4: 305–329.

Part III

Cloud-enabling geoscience projects

This part first introduces how to use cloud-enabled GIS using ArcGIS in the Cloud as an example (Chapter 7); and then demonstrates how to cloud-enable complex geoscience applications using three practical examples: (1) the GEOSS Clearinghouse project is used to demonstrate how to cloud-enable databases, spatial index, and spatial Web portal technologies (Chapter 8); (2) the Climate@Home project is used to demonstrate how to cloud-enable geoscience model simulations (Chapter 9); and (3) the Dust Storm Forecasting project is used to demonstrate how to leverage elastic cloud resources to support spiking computing incurred by a disruptive event (Chapter 10).

Chapter 7

ArcGIS in the cloud

Manzhu Yu, Pinde Fu, Nanyin Zhou, and Jizhe Xia

This chapter takes the ArcGIS[1] product suite in the cloud as a practical example to demonstrate how to use a cloud-enabled Geographic Information System (GIS). With practical use cases introduced, readers can obtain a clear understanding of the current status of cloud-enabled GIS.

7.1 INTRODUCTION

7.1.1 Why a geographical information system needs the cloud

Many 21st century challenges require the timely integration of vast amounts of geospatial information through a Geographic Information System (GIS) to address global and regional issues such as emergency response and planning. The timely integration of large amounts of information requires the readiness of a computing infrastructure with the following characteristics: sufficient computing capabilities, minimized energy cost, fast response, and wide accessibility to the public. Traditional GIS software, such as ArcGIS, is operated on desktops and local servers, and is focused on single users, mostly GIS experts. It requires the installation and maintenance of both hardware and software, and lacks the ability to support the needs of large-scale concurrent access. Cloud computing provides the computing capability to build and deploy GIS as a service, which can be referred to as *cloud-enabled GIS* or *Cloud GIS* (Mann 2011). The emergence of Cloud GIS is motivated by the need to establish new maintenance, usage, and billing modes which can address the existing computing issues and satisfy a broader user base. The technologies and architecture that cloud computing can offer are key areas of research and development for GIS solutions (Kouyoumjian 2010). New GIS solutions, such as the ArcGIS Online (introduced in Section 7.2.1), have been provided by delivering GIS as cloud

---

[1] See Esri at http://www.esri.com/software/arcgis.

services (Bhat et al. 2011). Cloud GIS can provide users with (1) no installation and maintenance of software, (2) unlimited computing resources and storage space, and (3) on-demand services. Additionally, Cloud GIS extends the application scope of GIS from geography to various social and business fields.

7.1.2 GIS examples that need the cloud

Cloud GIS can better support large-scale concurrent access and computing requirements, such as assisting emergency response. Cloud GIS has met the demands of emergency management by providing timeliness, interactivity, accessibility, and collaboration. For example, in 2010, when extensive flooding covered three quarters of the state of Queensland, Australia, responding agencies needed quick access to information for the rapidly changing situation. Esri Australia set up a GIS-driven common operating picture (COP) application in 12 hours that gave access to the latest, most accurate information on the situation in Brisbane. Another successful scenario is that Cloud GIS supports various social and business fields at an enterprise level to create solutions that help clients, including risk managers in organizations such as insurance and financial service companies. For example, ArcGIS Online helped the Wall Street Network deploy their clients' business continuity and disaster response plans even after the outage caused by flooding from Superstorm Sandy (Richardson 2012).

7.2 ArcGIS IN THE CLOUD

ArcGIS in the cloud currently includes ArcGIS Online, ArcGIS for Server, GIS Software as a Service, and the Mobile GIS service. ArcGIS Online is a preconfigured platform that can be accessed through a Web browser. ArcGIS for Server is a GIS server that can be deployed on cloud services. GIS Software as a Service is a cloud-based software that can be used as a service. Mobile GIS service is GIS Software as a Service for mobile operations.

7.2.1 ArcGIS Online

ArcGIS Online[1] is a cloud-based system for Web mapping and geographic information management, and is expanding and transforming into a complete Software as a Service (SaaS) application. ArcGIS Online provides the opportunity to gain insight into data quickly and easily without installing and configuring GIS software. ArcGIS Online offers GIS as a service including intuitive tools to create and publish maps and applications on demand, a rich collection of base maps, demographic maps, image services,

---

[1] See ArcGIS at http://qaext-ds.arcgis.com/about/.

and other data. It is convenient to share maps customized in ArcGIS Online through blogs, Web pages, applications, and Facebook or Twitter. Users can create, host, and share data and tiled map services using ArcGIS Online supported by the secure cloud.

7.2.1.1 Functionalities

Users can create Web maps using ArcGIS Online map services and geoservices, build custom Web and mobile applications using ArcGIS Web Mapping APIs and ArcGIS Mobile Runtime Software Development Kits (SDKs), and add the created maps into ArcGIS for Desktop. There are two main Web service options: Hosted Services and Content Services (Georelated 2012).

1. *Hosted Services*—Hosted services are ArcGIS Server services hosted and managed by Esri as part of the ArcGIS.com service. There are two types of hosted services available:
   - *Hosted Feature Service*—The service supports the query and edit of features as vectors.
   - *Hosted Tile Map Service*—The service provides pregenerated (cached) tiled map images. It also enables users to upload data and generate map tile images and host the data.
2. *Content Services*:
   - *Map Services*—Map services provide reference data services using Esri hosted data. SOAP, REST, and Rich Internet Application (RIA) controls/mapping application APIs:
      i. *Topographic Map Services*—These provide base mapping, demographic, reference, and others.
      ii. *Image Services*—These provide global land survey data as tiles.
      iii. *Bing Maps*—Bing base maps.
   - *Task Services*—Task services are accessible through SOAP and REST in RIA controls/mapping application APIs:
      i. *Geosearch*—The geosearch service provides searches for features or points of interest to support locating a map.
      ii. *Geocoding*—The geocoding service provides both forward and reverse geocoding for addresses, points of interest, and administrative locations. It supports transactional tasks and batches. Batches can be *Generated* or *Published* either returning their results or making them directly available as a map layer. Coverage supports the United States, Canada, and Europe based on Tele Atlas data.
      iii. *Routing*—The routing service provides routes and instructions with support for multiple languages. The coverage includes the United States, Canada, and Europe. The routing service is based on Tele Atlas data.

iv. *Geometry*—It is commonly used in geometric operations, such as Change CRS, Simplify, Buffer, Area, Length, Label Point, Convex Hull, Cut, Densify, Relation, Autocomplete, Cut, Difference, Generalize, Intersect, Offset, Reshape, Trim/Extend, and Union.

7.2.2 ArcGIS for Server

ArcGIS for Server[1] is a GIS server that can be deployed on Amazon EC2 and the Virtual Computing Environment (VCE) Vblock, which allows users to host GIS resources on their ArcGIS for Server systems. It also allows client applications such as Web mapping applications and mobile devices to use and interact with the resources. ArcGIS for Server supports various OS platforms, including Microsoft Windows Server, Red Hat Enterprise Linux AS/ES, SUSE Linux Enterprise Server, and Ubuntu. There are two important components related to ArcGIS for Server: ArcGIS Server AMIs and ArcGIS Server Cloud Builder.

ArcGIS Server AMIs are preconfigured AMI templates. Through these AMIs, users can quickly set up a fully functional ArcGIS Server instance on Amazon EC2. Existing ArcGIS for Server users can request access to ArcGIS Server AMIs through Esri Customer Service, by providing their Amazon EC2 Account ID. Then ArcGIS Server instances can be managed like any other Amazon EC2 instances.

ArcGIS Server Cloud Builder, a downloadable desktop application, gives users various options to launch ArcGIS for Server sites on Amazon EC2. It enables users to scale sites dynamically based on demands and backup their ArcGIS for Server sites (Chappell 2010).

7.2.2.1 Functionalities

- *GIS Services central management and delivery*—Each user has his/her own ArcGIS for Server site, where he/she can centralize the management of all the services for mapping, imagery, globes, geocoding, geodata management, and more. GIS functionalities, such as Web editing, network analysis, and schematics, as well as modeling, statistics, and other geo-analytic tools can also be provided.
- *On-demand response for maps and GIS tools*—ArcGIS for Server's architecture enables users to quickly scale their GIS systems to accommodate spikes in demand by adding and removing GIS servers as needed within their own infrastructure, in virtualized environments, and in the cloud.
- *Integration with enterprise applications*—The services created and managed with ArcGIS for Server can be integrated with enterprise

---

[1] See Esri at http://www.esri.com/software/arcgis/arcgisserver.

applications—giving the end-user the power of GIS with no extra efforts. Developers have access to the comprehensive Web mapping APIs and mobile runtime SDKs that can be used to create amazing mapping applications or integrate maps and GIS capabilities into existing Web, mobile, and desktop applications.

7.2.3 GIS software as a service

GIS Software as a Service provides centralized, cloud-based clients and applications that can easily solve complex problems using GIS tools and data. It integrates large volumes of geographic information and makes them available through easy-to-use Web applications. These applications are closely associated with both domain areas (e.g., geobusiness, policy, and community mapping) and the GIS community in general. The emergence of GIS Software as a Service is to meet users' demand of using GIS in a service-based environment without acquiring the data or the technology.

7.2.3.1 Functionalities

- *ArcLogistics*[1] is a cloud application for creating optimized routes in smaller amounts of time. This application is built on Amazon Web Services (AWS), with logic running in an Amazon EC2 VM and data stored in S3 (Amazon Simple Storage Service). The ArcLogistics client is not a Web browser, but a customized Windows application built with a Windows Presentation Foundation. Users can enter the number of vehicles and the necessary stops, and then get back optimized routing for these vehicles.
- *Business Analyst Online*[2] is a cloud application for working with demographic data, user data, and other information. By providing demographic information and other data, along with tools for working with that data, this cloud application can help business owners, planners, realtors, and others make better decisions.

  Business Analyst Online can be accessed from an ordinary Web browser. The application also exposes a Web service interface that can be used by other clients. For example, a corresponding iPhone application is provided so that third parties can also access these services. This application runs in Esri's own Internet-accessible data centers instead of AWS or other cloud services.
- *Community Analyst*[3] is a Web-based system that allows users to view and analyze demographic, public, and third-party sources of data

---

[1] See Esri, ArcLogistics Suite at http://www.esri.com/software/arclogistics-suite.
[2] See Esri, Business Analyst Online at http://www.esri.com/software/bao.
[3] See Esri, Community Analyst at http://www.esri.com/software/arcgis/community-analyst.

to better understand the overall community and make better policy decisions. One distinguishing feature is that Community Analyst uses up-to-date data, so that it can get information immediately for the exact area that users need—including standard geographies (down to the Census block group level), hand-drawn shapes, rings, or drive times around a location.

Community Analyst has proven efficient and significantly helpful. For example, the Epidemiology and Program Evaluation Branch of the Riverside County Department of Public Health made maps for each related staff member in the department (ArcNews 2011/2012). These maps are used to promote and protect the health of county residents and visitors, as well as ensure that services promoting the well-being of the community are available. According to the outcome of the maps, Community Analyst provides satisfying performance and insight for users.

7.2.4 Mobile GIS service

Mobile devices are an important user device for Cloud GIS, serving as a terminal for displaying, editing, collecting, and analyzing geospatial data. The growing popularity and wide use of mobile devices has enlarged the application scope of Cloud GIS. Through mobile applications, such as ArcGIS for Windows Mobile and Windows Tablet, and ArcGIS for Smartphones and Tablets, individual users and organizations can reach the functions provided by the server (e.g., ArcGIS for Server). Mobile GIS extends GIS and allows organizations to make accurate, real-time business decisions and collaborate in both field and office environments. It improves efficiency and accuracy of field operations, provides rapid data collection and seamless data integration, replaces paper-based workflows, and helps make timely and informed decisions.

For mobile workers, the cloud offers better mobility to improve workflow productivity and collaboration. Shared data and applications in the cloud can be immediately accessed to discover, view, edit, and save changes and to invoke geoprocessing functions for on-demand results.

7.2.5 Section summary

This session introduces four products of ArcGIS in the cloud: ArcGIS Online, ArcGIS for Server on the cloud, GIS Software as a Service, and mobile GIS service. Powerful GIS functions and services are supported by various Cloud GIS products and are delivered to users through both wired and wireless networks. These products provide GIS functions and services that can be accessed quickly and easily with nothing to install or set up, which is critical to the Internet and mobile era. The next section introduces three use cases using different products of the ArcGIS in the cloud. These use cases demonstrate in detail how the cloud could better support GIS and how these GIS products could serve users and their applications.

7.3 USE CASES

7.3.1 Regional analysis of Oregon using ArcGIS Online

Nowadays, urban planning and management has become an important application domain of GIS. Governments, especially city planning administrators, need macro control over urban and regional construction in order to make the city more livable and beautiful, as well as to come up with the best solutions when emergencies happen.

The map in this use case shows the regional analysis of a part of Oregon State, combining the urban area growth with the related infrastructure, natural elements, and other information (Figure 7.1). For example, the urban area in part of Oregon State in 2000 is marked in yellow, and in 2010 it is marked with both yellow and green (see Color Insert). From this map we can clearly see the urban growth of this area. In addition, different symbols represent different elements. For example, brown lines represent the primary roads; the shape, which looks like a tree, represents Park & Ride. The following steps illustrate how to create, save, and share the map made on ArcGIS Online.

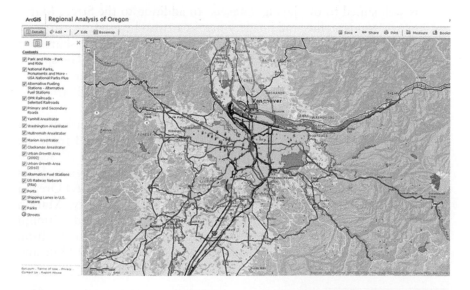

Figure 7.1 (See color insert.) Regional Analysis of Oregon map.

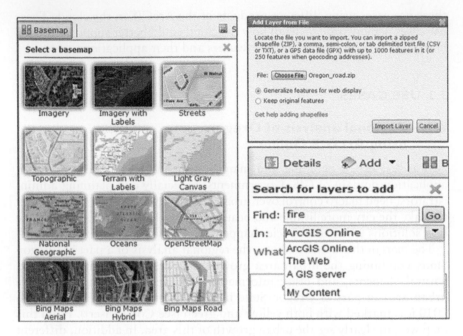

Figure 7.2 (See color insert.) Base maps and data offered by ArcGIS Online.

Step 1. Choose base maps—Make sure the base map is chosen based on the functions and features of the map. The backdrop for the data is the Street Map (Figure. 7.2), one of the ArcGIS Online base maps that is well suited to overlaying datasets. In addition to the Street Maps, ocean, topographic, and imagery maps can also be used.

Step 2. Add layers—Layers can either be those that have been shared on the Web, for example, through ArcGIS for Server or Open Geospatial Consortium, Inc., Web Map Services, KML or CSV layers, or personal data stored as delimited text files, in GPS Exchange Format, or as shapefiles.

Step 3. Add data—The core data used in this case are urban area distributions of Oregon in both 2000 and 2010 extracted from the U.S. Census Bureau.[1] The steps for retrieving the data are: (1) Search for *Roads* on the shapefiles 2010 page and choose Oregon from Primary and Secondary Roads; (2) Search for *Water* and select Oregon from Area Hydrography for counties: Clackamas County, Marion County, Multnomah County, Washington County, and Yamhill County; (3) Search for *Urban Growth Area* and select Oregon from Urban Growth Area (2010) and Urban Growth Area (2000); and (4) Download all the files that are chosen.

---

[1] See U.S. Census Bureau at http://www.census.gov/cgi-bin/geo/shapefiles2010/main.

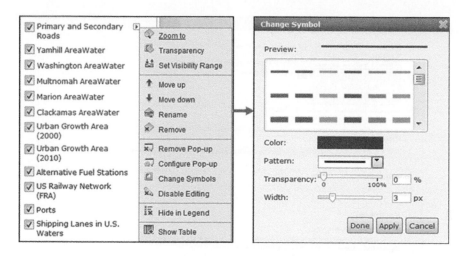

Figure 7.3 (See color insert.) Symbol modification of the Primary and Secondary Roads.

In order to perform an extensive analysis, related environmental data are also required and can be found in online GIS datasets, such as railway networks, Park & Rides, national parks, and fuel stations.

Step 4. Symbol modification—This step is to modify the symbols of the existing layers for better display. Users can change the symbol of the Primary and Secondary Roads into a brown and solid line with a width of 3px (Figure 7.3), and the symbol of different Area Water layers with various blue colors. They can also display the Urban Growth Area 2000 layer in yellow and the Urban Growth Area 2010 layer in green. Set all the other layers as the default. (See Color Insert.)

Step 5. Save the map—After creating the map, users can save it and add a description and attach tags to the map. A detailed description and attached tags provide a context, and help others find your map. The map is stored in My Content of the users' Esri account. Users can choose whether or not to share it on the Web with the public (Figure 7.4).

Step 6. Post and share—ArcGIS Online Sharing enables users to post and share their geospatial maps, layers, and tools among the ArcGIS community, or create a private group to exchange content related to a specific project or common activity.

This use case demonstrates how ArcGIS Online is ready to be used and delivered as a service, and introduces the intuitive tools and a rich collection of base maps and data to create and publish maps and applications on demand.

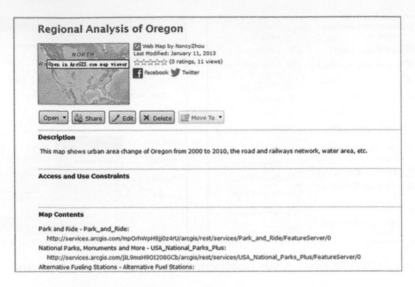

Figure 7.4 Map stored in My Content of an Esri account.

7.3.2 Use cases of ArcGIS for Server

7.3.2.1 Brisbane City Council Flood Common Operating Picture

Brisbane City Council (BCC) Flood Common Operating Picture (COP), built with the ArcGIS Server on Amazon EC2, was developed by the staff working in Esri Australia Pty. Ltd.[1] This system helps agencies respond to floods, and helps residents of affected areas access current information on the evolving situation. Similar works were conducted during many disasters in 2010—including the Deep Horizon oil spill in the Gulf of Mexico and earthquakes off the coast of the Maule region of Chile and near Port-au-Prince, Haiti. GIS and IT professionals quickly built a Web mapping GIS application in response to an overwhelming demand for information about each event. Access to these Web sites has demonstrated that, besides emergency responders and government officials, the public is also interested in timely, accurate, interactive maps. This situation reveals the requirements of concurrent access and computing processing capabilities.

The BCC Flood Map compiled flood data from across disaster-struck Brisbane—such as flood peaks, road closures, and evacuation centers—onto a map to provide a comprehensive, real-time scenario of the flood. An important aspect of the BCC Flood Map is a function which allowed users to turn on and off the information layers as needed, such as property damage

---

[1] See Esri, ArcWatch at http://www.esri.com/news/arcwatch/0611/feature2.html.

and evacuation center locations. Throughout the floods, Esri Australia's team continuously updated the BCC Flood Map with information fed from local and state government officials and emergency crews on the ground.

Cloud GIS is the key behind this application's success, because it is a tightly integrated system of components and content that can be deployed off premises within a matter of hours, and can be administered remotely. When a major emergency occurs, GIS professionals would not have the time or resources to pull existing equipment and software offline and repurpose them to support a single Web application serving tens of thousands of visitors daily. What they need is the access to a standby machine that is configured and ready to go within a few minutes. They also need massive enterprise resources that can instantaneously handle spikes in demand but also automatically scale down as Web traffic decreases.

This use case is one of many successful stories of Cloud GIS benefiting emergency management. These kinds of applications or services have efficiently resolved the demand for transmitting information about each emergency to the people involved in a short period of time.

7.3.2.2 Pennsylvania State Parks Viewer

Pennsylvania State Parks Viewer is a new interactive map of the Pennsylvania Department of Conservation and Natural Resources for discovering Pennsylvania State Park information. Like many other applications, it provides different base maps, such as the National Geographic World Map, World Street Map, and Aerial Imagery Map. There are also layers offering other geographic information, such as state boundaries, tourism regions, county boundaries, and watersheds.

In this application there are three modes to switch between: State Parks, State Forests, and Geology. Each mode provides different kinds of information about the parks and forests.

If parents want to take their kids camping or picnicking, they can choose the State Parks search (Figure 7.5) and select several catalogs, such as Camping, Education Program, Picnicking, and Sightseeing. Then they can get several results of the recommended parks, with their names, area, and location. If one of them is chosen, the detailed information (Figure 7.6) will appear in a large-scale map, along with directions from the starting location.

This application also has some useful tools for users to easily extract data, add data to ArcMap and Google Earth, create PDFs, and print. Other functions include distance measure, latitude/longitude coordinates, annotation, and weather forecast. Users can also make bookmarks, legends, and query for driving directions.

This use case demonstrates that besides supporting emergency management, Cloud GIS can also assist people to plan leisure time activities with convenience.

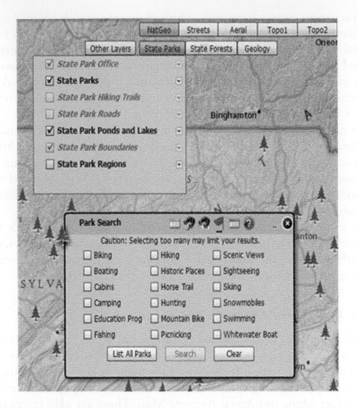

Figure 7.5 Park search in State Parks mode.

7.3.3 Section summary

This section presents three use cases using ArcGIS Online and ArcGIS for Server. These use cases demonstrate how cloud-enabled GIS can (1) leverage virtualized computing resources to provide Web services (e.g., map services and geoservices), (2) handle spikes in demand, and (3) provide preinstalled and configured virtual machines so that it is ready to use within a few minutes.

7.4 SUMMARY

This chapter provides several practical examples of Cloud GIS. Section 7.2 introduces ArcGIS suite products in the cloud: ArcGIS Online, ArcGIS for Server, GIS Software as a Service, and Mobile GIS service. Section 7.3 illustrates three use cases to further introduce different aspects of Cloud GIS,

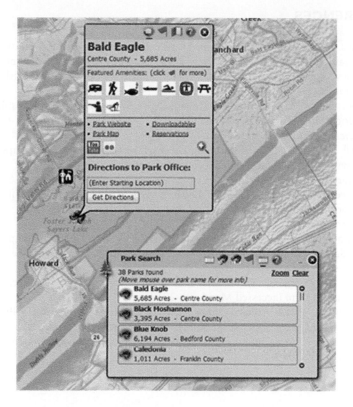

Figure 7.6 Detailed information for one searched site.

such as regional analysis, emergency management, and recommending tools for people's recreational life. These practical examples provide a firsthand experience about how to use Cloud GIS, and how to establish potential GIS solutions for challenges of the 21st century.

7.5 PROBLEMS

1. What applications can benefit from Cloud GIS?
2. What are the four components of ArcGIS in the cloud? What makes them exclusive?
3. Do you think ArcGIS Online is already an SaaS application? Why?
4. What are the ArcGIS for Server functionalities that make it different from local servers?
5. What are the functionalities of ArcGIS Online that make it different from ArcGIS Desktop?

REFERENCES

ArcNews. 2011/2012. Opening Up Health Reform: Riverside County's Public Health Department Takes a New Look at Grant Applications. Winter. http://www.esri.com/news/arcnews/winter1112articles/opening-up-health-reform.html (accessed March 29, 2013).

Bhat, M. A., R. M. Shah, and B. Ahmad. 2011. Cloud computing: A solution to geographical information systems (GIS). *International Journal on Computer Science and Engineering* 3, no. 2: 594–600.

Chappell D. 2010. GIS in the Cloud: The Esri Example. Esri White Paper. http://www.esri.com/library/whitepapers/pdfs/gis-in-the-cloud-chappell.pdf (accessed March 29, 2013).

Georelated.com. 2012. Geocloud: ESRI ArcGIS Online Reviewed. http://www.georelated.com/2012/10/cloud-web-mapping-service-api-review.html (accessed March 29, 2013).

Kouyoumjian, V. 2010. GIS in the Cloud: The New Age of Cloud Computing and GIS. Esri White Paper. http://www.esri.com/library/ebooks/gis-in-the-cloud.pdf (accessed March 29, 2013).

Mann, K. 2011 (Spring). Cloud GIS. Esri. http://www.esri.com/news/arcuser/0311/cloud-gis.html (accessed March 29, 2013).

Richardson, K. 2012. ArcGIS Online Helps Wall Street Network Do Business in the Cloud after Superstorm Sandy: From Reactive to Proactive. *ArcWatch: GIS News, Views, and Insights*. http://www.esri.com/news/arcwatch/ (accessed March 29, 2013).

Chapter 8

Cloud-enabling GEOSS Clearinghouse

Kai Liu, Douglas Nebert, Qunying Huang,
Jizhe Xia, and Zhenlong Li

GEOSS (Global Earth Observation System of Systems)[1] Clearinghouse (CLH) is a comprehensive data management system, and is the core component of GEOSS that provides harvesting and searching capabilities for the geospatial components and services registered in the GEOSS. This chapter introduces how to cloud-enable CLH with the deployment and configurations needed for running the CLH in Amazon EC2 cloud service.

8.1 GEOSS CLEARINGHOUSE: BACKGROUND AND CHALLENGES

8.1.1 Background

A variety of Earth Observation (EO) systems monitor the planet Earth and generate large amounts of EO data on a daily basis. The observations provide baseline data about the Earth with specific time snapshots, trajectories, and events. These data are useful to support different Societal Benefit Areas (SBAs) including agriculture, biodiversity, climate, disasters, ecosystems, energy, health, water, and weather defined by the intergovernmental Group on Earth Observation (GEO) (GEO 2009). Recognizing the value of EO data and global dissemination requirements, the GEO was established to build a GEOSS Common Infrastructure (GCI) to help manage, integrate, access, and share the global EO data to address worldwide and regional problems (Christian 2005).

CLH is the engine that drives the entire GCI. It was developed and maintained by the Center of Intelligent Spatial Computing for Water/ Energy Science (CISC) at George Mason University, in collaboration with the U.S. Geological Survey (USGS). CLH builds a bridge between the EO data providers and GEOSS end users by (1) collecting the metadata of EO data from multiple data sources through harvesting mechanisms, and

---

[1] See Group on Earth Observations at http://www.earthobservations.org/index.shtml.

(2) providing local and remote search capabilities to enable users to discover EO data (Liu et al. 2011).

8.1.2 Challenges

As an exemplification of spatial Web portals (Yang et al. 2007), CLH is designed to support worldwide access to the spatial data resources. Several challenges are considered when developing CLH:

- *Big data*—Big data is a tough challenge in CLH with respect to the three Vs:
  1. *Volume*—GEOSS is a worldwide system containing a large amount of metadata for the vast amounts of EO data collected from a wide distribution of catalogs in GCI. Once EO data providers register their data catalogs or services in the GEOSS registry, CLH will harvest these metadata continually with a time interval (e.g., one hour or one day) depending on the update frequency of the data sources. By January 2013, 104 catalogs and 120,000 metadata records were registered into CLH. These data are big and growing rapidly.
  2. *Velocity*—Frequent updating of the datasets and metadata.
  3. *Variety*—The metadata formats/standards (FGDC CSDGM[1] Metadata Standards, Dublin core,[2] and ISO-19139[3]) and access protocols, for example, Catalog Service for Web (CSW),[4] Search/Retrieval via URL (SRU),[5] and Really Simple Syndication (RSS),[6] are different.
- *Spatiotemporal search and full text search*—The search capability is the most important function for CLH. However, it is not easy to provide high-performance search functions in CLH because the data volume is massive and each record contains both spatiotemporal and text information. Combining spatiotemporal search and full text search will dramatically increase computational complexity. The metadata in CLH are in fact spatiotemporal metadata, which not only store the literal description of EO data but also are associated with spatiotemporal information. To discover applicable data from the geospatial metadata in CLH, users need to use *What* to find the description information, *Where* to locate the geographic extend, and *When* to specify the temporal information.

---

[1] See Federal Geographic Data Committee at http://www.fgdc.gov/metadata/csdgm/.
[2] See Dublin Core Metadata Initiative at http://dublincore.org/.
[3] See ISO at http://www.iso.org/iso/catalogue_detail.htm?csnumber=32557.
[4] See OGC at http://www.opengeospatial.org/standards/cat.
[5] See SRU at http://www.loc.gov/standards/sru/specs/transport.html.
[6] See RSS 2.0 Specification at http://feed2.w3.org/docs/rss2.html.

- *Concurrent access*—Recent developments in distributed geographic information processing (Yang and Raskin 2009) and the popularization of Web and wireless devices enable a massive number of end users to access geospatial systems concurrently (Goodchild et al. 2007). Being accessed all over the world, CLH has massive concurrent access over time. The large concurrent requests impose heavy stress for CLH on both hardware and software, which may result in long response time to users. If the response time is longer than two seconds, the users may feel frustrated (Nah 2004). As an operational application that is accessed by users worldwide, it is particularly important for CLH to provide high performance data access capability.

To tackle these challenges, Amazon Web Services (AWS) is used to leverage various advantages of cloud services, such as the Amazon EC2 with scalable computing power and the Amazon stable Elastic Block Store (EBS) with high I/O and the scalable mechanism. Through that, CLH can effectively handle concurrent access at high peak hours. These advantages will be introduced as special considerations in Section 8.2.2.

8.2 DEPLOYMENT AND OPTIMIZATION

8.2.1 General deployment workflow

Figure 8.1 demonstrates how to deploy CLH onto Amazon EC2, where the boxes in grey indicate the steps requiring special consideration. The cloud consumers need to have an account to deploy applications to EC2. The steps for creating an account can be referred to in Chapter 4, Section 4.3.1. After signing up for the account, the general steps are as follows:

- *Step 1. Authorize network access*—As described in Chapter 4, Section 4.3.1, this step is to authorize the network access by opening the appropriate ports. In this case, port 22 is opened to enable the communication between a local server and EC2 instances, and port 80 is opened to enable the access of CLH through the Web browser.
- *Step 2. Launch an instance*—CLH is a Web application that is hosted by Tomcat and runs on the Linux machine with a spatial database. Therefore, a public Amazon Machine Image (AMI) with PostgreSQL 8.4 and PostGIS 1.5 is selected to launch an Amazon EC2 instance.
- *Steps 3, 4, and 6.* Steps 3, 4, 5, and 6 in Figure 8.1 are used to customize the instance. Steps 3, 4, and 6 are optional, but they make the system more reliable with more storage. These steps are covered in Section 8.2.2.1.

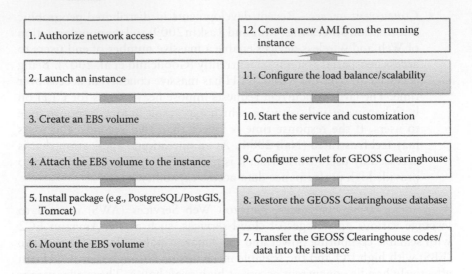

Figure 8.1 The process of deploying CLH onto Amazon EC2 (gray boxes indicate the steps that require special considerations).

Table 8.1 Compatibility Matrix among Different Versions of PostgreSQL and PostGIS*

PostgreSQL Version	PostGIS 1.3	PostGIS 1.4	PostGIS 1.5	PostGIS 2.0 Trunk
8.4	Yes	Yes	Yes	Yes
9.0	No	No	Yes	Yes
9.1	No	No	Yes	Yes

*See PostGIS at http://trac.osgeo.org/postgis/wiki/UsersWikiPostgreSQLPostGIS.

- *Step 5. Install package*—In Step 5, PostgreSQL/PostGIS and Tomcat should be installed and configured. PostgreSQL/PostGIS is the database software, which stores the spatial metadata records of CLH, and Tomcat is a servlet container that is used to host CLH. Before installing PostgreSQL/PostGIS, the dependent packages, such as Proj (cartographic projections library),[1] GEOS (Geometry Engine-Open Source),[2] and libxml (XML C parser and toolkit),[3] should be installed. One important aspect consumers should be aware of is that there are incompatibility issues among different versions of PostgreSQL and PostGIS (Table 8.1), as well as PostGIS and GEOS (Table 8.2). In this case, a Ubuntu AMI with PostgreSQL 8.4 and PostGIS 1.5 is used to launch the instance. Hence, it saves time for consumers to install the packages.

---

[1] See Proj. 4 at http://trac.osgeo.org/proj/.
[2] See GEOS at http://trac.osgeo.org/geos/.
[3] See the XML C parser and toolkit of Gnome at http://www.xmlsoft.org/.

Table 8.2 Compatibility Matrix among Versions of GEOS and PostGIS

GEOS version	PostGIS 1.3	PostGIS 1.4	PostGIS 1.5	PostGIS 2.0 Trunk
None	Yes (not recommended)	No	No	No
2.2	Yes (not recommended)	No	No	No
3.0	No	Yes (not recommended)	Yes	Yes
3.1	No	Yes (not recommended)	Yes	Yes
3.2	Yes	Yes	Yes	Yes (not recommended)
3.3	Yes	Yes	Yes	Yes

After configuring the database, consumers can set up the Tomcat with the following commands:

```
root@ip-10-189-149-104:/root$ cd /opt
root@ip-10-189-149-104:/opt$ wget http://www.
  poolsaboveground.com/apache/tomcat/tomcat-6/v6.0.26/
  bin/apache-tomcat-6.0.26.tar.gz
root@ip-10-189-149-104:/opt$ tar -zxvf apache-
tomcat-6.0.26.tar.gz
```

In order to make Tomcat work properly, the Java Development Kit (JDK) is also required to be installed and configured (Chapter 5).

- *Step 7. Transfer the CLH code and data*—How to transfer a massive data volume to the cloud is detailed in Chapter 5, Section 5.3.
- *Step 8. Restore the database*—After transferring the CLH applications to the instance and restoring data to the database for the instance, the following codes can be used to restore the database. The first line is used to change the owner of a database dump file (geoss. sql) to PostgreSQL; the second line is used to switch to a PostgreSQL user; the third line is used to create a new database: geoss; the fourth line is used to restore the database.

```
root@ip-10-189-149-104:/mnt$ chown postgres:postgres
  geoss.sql
root@ip-10-189-149-104:/mnt$ su postgres
bash-3.2$ createdb geonetwork
bash-3.2$ psql geonetwork < geoss.sql
```

- *Step 9. Configure servlet for CLH*—A virtual user, "Tomcat" is created to run the Tomcat in this step. The following codes show the

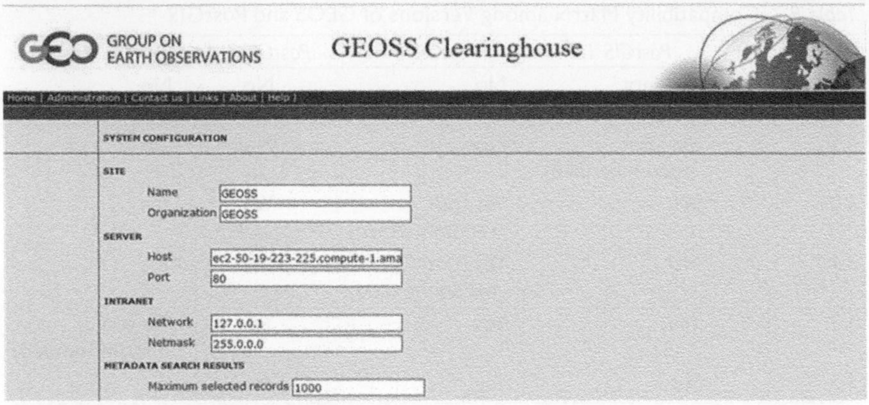

Figure 8.2 Configuring the CLH.

configuration for the `/etc/rc.local` file to enable CLH to start automatically when the system boots up.

```
sudo -u tomcat /opt/tomcat/bin/startup.sh
```

- *Step 10. Start the service and customize the application*—If the database and Tomcat is configured properly, users should be able to access CLH through a Web browser after Tomcat is started. However, additional customizations are necessary, including setting the URL for remote search (detailed in Section 8.3.2) and changing the password for the administrator. Figure 8.2 shows the interfaces of setting the URL for remote search. The configuration page is under administration-system configuration after login to CLH. Users need to change the server URL to the actual domain name provided by cloud providers.
- *Step 11. Configure the load balance or scalability*—This step is optional to deploy CLH onto the cloud service, but it would make the system more flexible and scalable. Load balance and scalability are introduced in Sections 8.2.2.2 and 8.2.2.3.
- *Step 12. Create an AMI from the running instance*—Finally, a new AMI can be created based on the running instance of CLH (refer to Chapter 4, Section 4.3 for how to create an AMI).

8.2.2 Special considerations

8.2.2.1 Data backup

The data, log files, and applications could be copied to a separate EBS volume for backup. Both the command line tool and the AWS Web

Management Console can be used to create and attach the EBS volume to the CLH instance. The EBS volume is mounted to the PostgreSQL directory. Such a separate volume has two benefits: (1) it can be used to restore CLH from the volume in case the current instance crashes, and (2) the volume could be any size ranging from 1 GB to 1 TB.

The following steps show how to create and use EBS volume.

- *Step 1*. Create an EBS volume from scratch with no content in the Web console and make sure the selected EBS volume zone is the same as the zone of the CLH instance (Figure 8.3).
- *Step 2*. Attach the volume to the running instance. Follow the convention of using/dev/sdf for the first mount point (Figure 8.3). If there are multiple volumes to be attached, follow the alphabet sequence for each volume, for example, /dev/sdg,/dev/sdh, etc.
- *Step 3*. Log in to the instance and move the PostgreSQL data partition to the EBS volume. After logging into the instance, consumers will: (1) make a file system on the new volume, through the mkfs command; (2) make sure the Tomcat and PostgreSQL services are shut down properly; (3) create a directory on the EBS Boot Volume for backup (e.g., /mnt/datavol _1); (4) mount the/dev/sdh volume at the new mount point; and finally (5) backup the data, log files, and application to the new location.

```
[root@ ip-10-189-149-104~] mkfs -t ext3 /dev/sdh#
  make a file system
[root@ip-10-189-149-104~] mkdir /mnt/datavol_1
[root@ip-10-189-149-104~] mount /dev/sdh
  /mnt/datavol_1/
```

The above command shows commands for creating and attaching an EBS volume (assuming the instance domain name is *ip-10-189-149-104*, and the user account is "root").

Figure 8.3 Creating a new EBS volume and attaching to the instance.

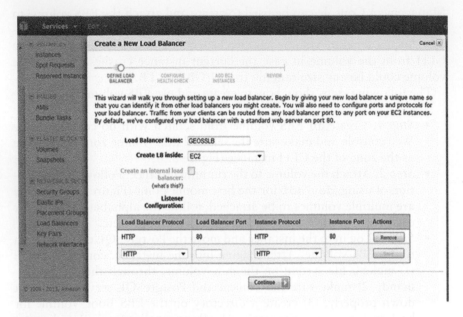

Figure 8.4 Configuring load balance service.

8.2.2.2 Load balancing

After successfully launching the Amazon EC2 instance, consumers can configure advanced services for the instance, such as load balance and scalability. AWS provides the load balancer[1] to balance the user request loads on multiple EC2 instances. Since CLH may expect to handle requests from massive numbers of users concurrently, the load balance service would be helpful to balance the traffic load of the system. Under the Network & Security section, users can click Load Balancers to start configuring the load balance service (Figure 8.4).

In this case, only one copy of the database is used while using load balance services, which means all instances will access the same database configured on the first instance.

8.2.2.3 Auto-scaling

Geospatial applications may experience hourly, daily, or weekly variability in user access, which requires various computing capabilities at different times to minimize the costs and to meet the performance requirements. Scalable EC2 services could facilitate geospatial applications by elastically satisfying computing requirements. AWS provides the auto-scalar[2]

---

[1] See Amazon Web Services at http://aws.amazon.com/elasticloadbalancing/.
[2] See Amazon Web Services at http://aws.amazon.com/autoscaling/.

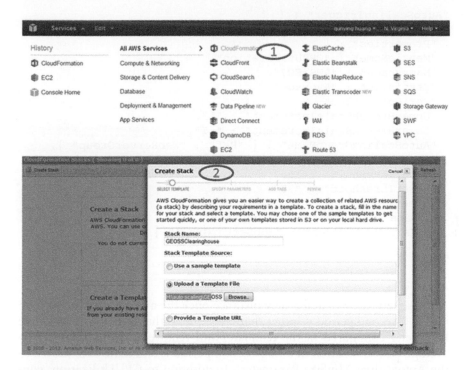

Figure 8.5 Using the CloudFormation service to configure auto-scaling capability through the Web console.

to implement such a functionality. The cloud consumers can also use the CloudFormation service through the Web console (Figure 8.5).

Consumers can also configure the auto-scaling service through the command line.

```
cfn-create-stack GEOSSClearinghouse --template-file
   GEOSSClearinghouse.template --region us-east-1
   --awsaccesskey=FAKEKEY --awssecretkey=FAKEKEY2
   --parameters="KeyName=GeoNet; InstanceType=m1.large"
```

However, no matter which services or tools are used to configure the auto-scaling function, consumers need to set up a template (i.e., the Stack Template Source in Figure 8.4 and the input for parameter template file when using the above-mentioned command `cfn-create-stack`). Appendix 8.1 is an example of the template. The template is a JSON[1] file that defines infrastructure requirements. In the template, the most important object required to define is the resource trigger. Within the sample template, the auto-scaling trigger is based on the CPU utilization of the Web servers as shown below.

---

[1] See Introducing JSON at http://www.json.org.

```
"AutoScalingTrigger" : {
  "Type" : "AWS::AutoScaling::Trigger",
  "Properties" : {
  "MetricName" : "CPUUtilization",
  "Namespace" : "AWS/EC2",
  "Statistic" : "Average",
  "Period" : "300",
  "UpperBreachScaleIncrement" : "1",
  "LowerBreachScaleIncrement" : "-1",
  "AutoScalingGroupName" : { "Ref" : "WebServerGroup" },
  "BreachDuration" : "600",
  "UpperThreshold" : "90",
  "LowerThreshold" : "75",
  "Dimensions" : [ {
    "Name" : "AutoScalingGroupName",
    "Value" : { "Ref" : "WebServerGroup" }
} ]
}
},
```

These settings will add a new instance once the average CPU utilization in a 300-second period is larger than 90% (up to the Autoscaling MaxSize Parameter configured in the template) and remove an instance once the average CPU utilization in a 300-second period is less than 75% (down to the AutoScaling MinSize Parameter). In addition to CPUUtilization, consumers can also use other values for the MetricName parameter, such as Latency and RequestCount. Latency will trigger the scaling of instances based on the Web service response latency time, and RequestCount is based on the number of user concurrent request numbers. More information about the template can be found at the AWS Web site.[1]

A snapshot for the EBS volume is created automatically every 5 minutes to store the PostgreSQL/PostGIS, CLH codes and temporary data, so that the newly scaled instance can attach the latest data volume with the synchronized data. The system will only keep one copy of the latest data volume (after a new EBS volume is created, it will replace the previous version) to save costs since Amazon charges storage fees based on the total volume size used.

8.2.3 The differences from the general steps in Chapter 5

CLH integrates several complex components, including database management (PostgreSQL/PostGIS), metadata resource management and harvest, and geospatial data preview. The potential computing intensity from big

---

[1] See Amazon Web Services at http://aws.amazon.com/cloudformation/aws-cloudformation-templates/.

data management and concurrent requests further contribute to the complexities of the entire system. During deployment of CLH onto Amazon EC2, we have used several special strategies to make it fully functional, including a backup of database, and configuring server load balancing and auto-scaling.

8.3 SYSTEM DEMONSTRATION

The cloud enabled CLH can be accessed at GEOSS Clearinghouse at http://ec2-50-19-223-225.computc-1.amazonaws.com/geonetwork. This section demonstrates how to discover geospatial datasets and resources using both local search (through CLH interface) and remote search (via standard Web service interfaces of CLH instead of Web GUI).

8.3.1 Local search

Local search function enables users to discover records based on title, keyword, spatial location, and temporal information. The results can be sorted by different rules (e.g., relevance, date, title, alphabetical order, popularity, and rate). Buttons for accessibility tools are dynamically added to each record panel according to the resource types (e.g., "Interactive Map" will be added to WMS). Users can also use integrated visualization tools (powered by OpenLayers[1]) to interact with data services. Figure 8.6 shows

Figure 8.6 (See color insert.) Search results for global "Annual Sum NDVI Annual Rainfall."

---

[1] See OpenLayers at http://openlayers.org/.

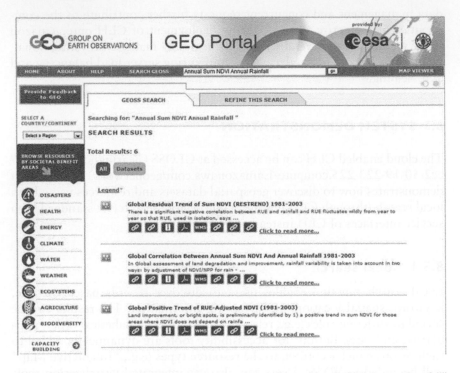

Figure 8.7 (See color insert.) Search results for global "Annual Sum NDVI Annual Rainfall" from the GEO Portal.

the results when searching global "Annual Sum NDVI Annual Rainfall" between 1980 and 2013.

8.3.2 Remote search

CLH supports various protocols for searching EO data remotely. It enables global users to customize their applications to use the EO records from GEOSS. By using the remote search capabilities, users could send search requests from their clients to CLH. The search protocols supported by CLH include CSW, SRU, and RSS. Figure 8.7 shows the CSW search results against CLH from the Web GUI of GEO Portal.

8.4 CONCLUSION

Cloud computing offers both economic and technical advantages to support CLH.

8.4.1 Economic advantages

One of the greatest advantages is that a traditional server is no longer required for hosting CLH. The monthly expenses on electricity, a cooling system, physical facilities, and system administrator for maintaining a server are saved. Table 8.3 shows the cost comparison between using AWS services and a local server (Huang et al. 2011). The fixed (one-time) cost and monthly

Table 8.3 Cost Comparison between AWS Services and a Local Server*

	Cost	Rent Model	Amazon EC2	Local Server
Fixed (One-Time) Cost			None	Around $2,000 to Purchase the Server
Monthly	Network cost		$1 for data transferring/ month	None (Included in the maintenance fee)
	Storage		$22/month	None (Included in the first purchase)
	Computing power	Not reserved	$255/month	~$17 (Assuming the server can be used for 10 years, and then the cost each month would be (2000/(10*12))
		Reserved	$57/month ($676 if reserving a large instance under heavy utilization for three years); $13/month ($1,028 if reserving for 3 years)	
	Maintenance cost (Cooling, system, sys admin, maintenance, room, etc.)		None	$200/per month (Assuming $100 for cooling, network, and room fee, and $100 for paying a sys admin to check and maintain the server)
Yearly	Total	Not reserved	$3,300	$2,604
		Reserved	$432 (instance reserved for 3 years)	

Note: The cloud computing cost is based on the standard large instance with two CPU cores of Amazon EC2 to host CLH. If more advanced instance types are selected and advance cloud services (e.g., scalability) are configured, the computing cost would increase.

*Compared on March 27, 2013.

cost in each year for a 10-year period are compared. The yearly cost for the AWS services is $432 when reserving a standard large instance with 2 CPU cores for a three-year commitment ($1,028 in total for the instance for three years at March 27, 2013); the yearly cost for a local server is $2,604. It saves about 83.6% per year when deploying and hosting CLH on cloud services.

8.4.2 Technical advantages

The cloud service (Amazon EC2) offers the scalability to maintain the performance of CLH when experiencing massive concurrent accesses. Amazon EC2 also offers a highly reliable environment for CLH because EC2 runs within Amazon's proven network infrastructure and data centers. The deployment of CLH to Amazon EC2 illustrates that cloud computing provides a cost-effective approach for hosting and operating geospatial applications in a highly scalable and reliable computing environment.

8.5 SUMMARY

This chapter introduces how to use cloud computing to support CLH. Section 8.1 introduces the background and challenges of CLH. Section 8.2 demonstrates cloud deployment of CLH. Section 8.3 illustrates the user scenarios of CLH, and Section 8.4 concludes the chapter by summarizing the economic and technical advantages of deploying CLH onto cloud services.

8.6 PROBLEMS

1. What are the general steps of deploying GEOSS Clearinghouse onto the cloud? What are the differences from the general steps in Chapter 5?
2. How do you migrate data to an attachable Amazon EBS volume?
3. Which cloud services can be used to balance the system load? Discuss how to use them.
4. What scalable services are provided by AWS? How do you use them?
5. Using GEOSS Clearinghouse as an example, explain the technical advantages of cloud-enabled geoscience applications.

APPENDIX 8.1 TEMPLATE FOR CREATING AN AUTO-SCALING FUNCTION

```
{
    "AWSTemplateFormatVersion" : "2010-09-09",
    "Description" : "Create a load balanced, auto scaled CLH
        Web site. ",
```

```
   "Parameters" : {
    "InstanceType" : {
     "Description" : "Type of EC2 instance to launch",
     "Type" : "String",
     "Default" : "m1.large"
    },
    "WebServerPort" : {
     "Description" : "The TCP port for the Web Server",
     "Type" : "String",
     "Default" : "80"
    },
    "KeyName" : {
     "Description" : "The EC2 Key Pair to allow SSH access
        to the instances",
     "Type" : "String"
    }
},


   "Mappings" : {
    "AWSInstanceType2Arch" : {
     "m1.large" : {"Arch" : "64"},
     "m1.xlarge" : {"Arch" : "64"}
    },
    "AWSRegionArch2AMI" : {
     "us-east-1" : {"64" : "xxxxx"},
     "us-west-1" : {"64" : " xxxxx "},
     "eu-west-1" : { "64" : " xxxxx "},
     "ap-southeast-1" : {"64" : " xxxxx "},
     "ap-northeast-1" : {"64" : " xxxxx "}
    }
},

"Resources" : {
  "WebServerGroup" : {
   "Type" : "AWS::AutoScaling::AutoScalingGroup",
   "Properties" : {
    "AvailabilityZones" : {"Fn::GetAZs" : ""},
    "LaunchConfigurationName" : {"Ref" : "LaunchConfig"},
    "MinSize" : "1",
    "MaxSize" : "3",
    "LoadBalancerNames" : [{"Ref" : "ElasticLoadBalancer"}]
  }
},

  "LaunchConfig" : {
   "Type" : "AWS::AutoScaling::LaunchConfiguration",
   "Properties" : {
    "KeyName" : {"Ref" : "KeyName"},
```

```
      "ImageId" : {"Fn::FindInMap" : ["AWSRegionArch2AMI",
                        {"Ref" : "AWS::Region"},
                          {"Fn::FindInMap" :
["AWSInstanceType2Arch", {"Ref" : "InstanceType"},
                             "Arch"]}]},
      "UserData" : {"Fn::Base64" : {"Ref" :
"WebServerPort"}},
      "SecurityGroups" : [{"Ref" : "InstanceSecurityGroup"}],
        "InstanceType" : {"Ref" : "InstanceType"}
    }
},

"CPUBasedTrigger" : {
  "Type" : "AWS::AutoScaling::Trigger",
  "Properties" : {
    "MetricName" : "CPUUtilization",
    "Namespace" : "AWS/EC2",
    "Statistic" : "Average",
    "Period" : "300",
    "UpperBreachScaleIncrement" : "1",
    "LowerBreachScaleIncrement" : "-1",
    "AutoScalingGroupName" : {"Ref" : "WebServerGroup"},
    "BreachDuration" : "600",
    "UpperThreshold" : "90",
    "LowerThreshold" : "75",
    "Dimensions" : [{
     "Name" : "AutoScalingGroupName",
     "Value" : {"Ref" : "WebServerGroup"}
    }]
  }
},

"ElasticLoadBalancer" : {
  "Type" : "AWS::ElasticLoadBalancing::LoadBalancer",
  "Properties" : {
    "AvailabilityZones" : {"Fn::GetAZs" : ""},
    "Listeners" : [{
    "LoadBalancerPort" : "80",
    "InstancePort" : {"Ref" : "WebServerPort"},
    "Protocol" : "HTTP"
    }],
    "HealthCheck" : {
    "Target" : {"Fn::Join" : ["", ["HTTP:", {"Ref" :
                  "WebServerPort"}, "/"]]},
    "HealthyThreshold" : "3",
    "UnhealthyThreshold" : "5",
    "Interval" : "30",
    "Timeout" : "5"
    }
```

```
      }
  },

  "InstanceSecurityGroup" : {
    "Type" : "AWS::EC2::SecurityGroup",
    "Properties" : {
    "GroupDescription" : "Enable SSH access and HTTP access
                          on the inbound port",
    "SecurityGroupIngress" : [{
     "IpProtocol" : "tcp",
     "FromPort" : "22",
     "ToPort" : "22",
     "CidrIp" : "0.0.0.0/0"
    },
    {
     "IpProtocol" : "tcp",
     "FromPort" : {"Ref" : "WebServerPort"},
     "ToPort" : {"Ref" : "WebServerPort"},
     "CidrIp" : "0.0.0.0/0"
    }]
    }
   }
  },

  "Outputs" : {
   "URL" : {
   "Description" : "The URL of the website",
   "Value" : {"Fn::Join" : ["", ["http://", {"Fn::GetAtt" :
                 ["ElasticLoadBalancer", "DNSName"]}]]}
  }
 }
}
```

REFERENCES

Christian, E. 2005. Planning for the Global Earth Observation System of Systems (GEOSS). *Space Policy* 21, no. 2: 105–109.

GEO. 2009–2011 Work Plan [online], 2009. http://www.earthobservations.org/documents/work%20plan/geo_wp0911_rev2_091210.pdf (accessed January 4, 2013).

Goodchild, M. F., M. Yuan, and T. J. Cova. 2007. Towards a general theory of geographic representation in GIS. *International Journal of Geographical Information Science* 21, no. 3: 239–260.

Huang, Q., D. Nebert, C. Yang, and K. Liu. 2011. GeoCloud Project Report—CLH [online]. http://www.fgdc.gov/initiatives/geoplatform/geocloud/reports/fgdc-geocloud-project-report-geonetwork.pdf (accessed March 4, 2013).

Liu, K., C. Yang, W. Li, Z. Li, H. Wu, A. Rezgui, and J. Xia. 2011. The CLH High Performance Search Engine. *The 19th International Conference on Geoinformatics*. June 24–26, Shanghai, China.

Nah, F. F. H., X. Tan, and S. H. Teh. 2004. An empirical investigation on end-users' acceptance of enterprise systems. *Information Resources Management Journal (IRMJ)* 17, 3: 32–53.

Yang, P., J. Evans, M. Cole, N. Alameh, S. Marley, and M. Bambacus. 2007. The emerging concepts and applications of the spatial Web portal. *Photogrammetric Engineering and Remote Sensing* 73, no. 6: 691.

Yang, C. and R. Raskin. 2009. Introduction to distributed geographic information processing research. *International Journal of Geographical Information Science* 23, no. 5: 553–560.

Chapter 9

Cloud-enabling Climate@Home

Jing Li, Zhenlong Li, Min Sun, Qunying Huang, Kai Liu, and Myra Bambacus

The Climate@Home project is an initiative that is designed to build a virtual supercomputer to support climate modeling. Cloud services are utilized to support the management of massive model outputs and computing resources and to facilitate efficient analysis of the outputs. This chapter introduces the deployment of the Climate@Home system onto cloud services.

9.1 CLIMATE@HOME: BACKGROUND AND CHALLENGES

9.1.1 Background

Climate modeling reconstructs the past and predicts future climate conditions through quantifying the interaction among the atmosphere, land, ocean, and sea ice with numerical descriptions (Skamarock and Klemp 2008). Climate modeling is featured with both computing intensity and data intensity. On one hand, climate models can generate massive datasets to describe climatological and geophysical dynamics. On the other hand, to avoid systematic errors caused by model configurations, multiple runs are necessary (Schmidt et al. 2006). Considerable amounts of computing resources are needed to finish these runs. As a pilot study, the Climate@Home project seeks an alternative approach to address the computation requirements of climate modeling and to contribute to the global climate studies. The primary objective of this project is to create a virtual supercomputer by leveraging citizen computing resources that can support multiple model simulations. The NASA Goddard Institute for Space Studies (GISS) ModelE[1] is chosen as the climate model for demonstration purposes.

Given the primary objective of the project, the core task is to develop a computing and data management system that serves as a comprehensive solution for integrating citizen contributed resources and facilitating scientific

---

[1] See NASA at http://www.giss.nasa.gov/tools/modelE/.

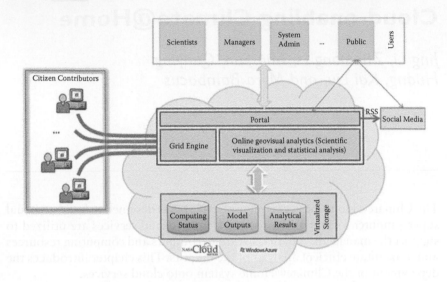

Figure 9.1 Climate@Home system architecture.

analysis. This system should have the following functions: (1) A collaborative computing environment where model simulations can be dispatched to computing devices; (2) A data warehouse that stores datasets from multiple sources including model outputs, analytical results, and resource status; (3) An online visual analytical component to facilitate the analysis of early scientific explorations and the status of the system; and (4) a set of secured online access interfaces to the above three functions. Figure 9.1 shows the architecture of Climate@Home. Two major components are: (1) a spatial Web portal (Yang et al. 2007) integrating multiple functions and (2) a BOINC[1] server, which serves as the grid engine that manages model runs and datasets.

- *Spatial Web portal*—Considering the complexity of the system architecture and multiple functions offered to different levels of users (e.g., scientists and volunteers), a spatial Web portal is developed to integrate all front-end components with back-end functions. The public and private interfaces of the portal allow scientists, technical staff, and resource contributors to access these components in a secured manner. To build the spatial Web portal, Drupal[2] is used as the container for integrating those components and functions. Drupal is an open-source content management system (CMS) that provides flexible interfaces for users to configure the site and adopt packages contributed by

---

[1] See BOINC at http://boinc.berkeley.edu/.
[2] See Drupal at http://drupal.org/.

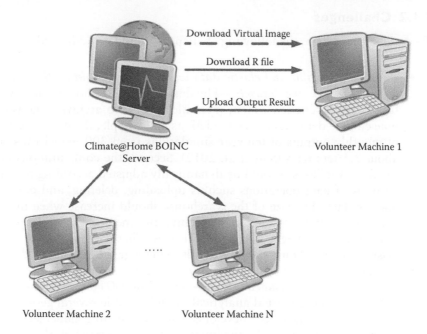

Download Virtual Image

Download R file

Upload Output Result

Climate@Home BOINC
Server

Volunteer Machine 1

Volunteer Machine 2

Volunteer Machine N

Figure 9.2 The BOINC-based Climate@Home simulation workflow.

community developers. A geovisual analytical portlet and a resource management portlet based on online Web mapping applications (Bing Maps[1] and Google Earth[2]) are integrated with the portal. These two portlets support the geovisual analysis of the climate data and visual management of contributed computing resources (Sun et al. 2012).

- The BOINC server is used as the grid engine, which is responsible for dispatching ModelE simulation tasks, collecting model outputs, and coordinating with the BOINC clients (Figure 9.2). Every computing device from citizen contributors is treated as a BOINC client. The BOINC can handle various platforms such as Windows, Linux/x86, and Mac OS/X (Anderson 2004). Since ModelE is a Linux-based climate model, the VirtualBox from Oracle (Anderson 2010) is chosen as the virtualization software to create Linux virtual machines on client devices. The Virtual Image is downloaded to the client from first-time citizen contributors. The BOINC server continuously dispatches model runs to the client. Once a model run is completed at the client side, model outputs are converted to NetCDF[3] data files and are uploaded to the BOINC server.

---

[1] See Bing Maps at http://www.bing.com/maps/.
[2] See Google Earth at http://www.google.com/earth/index.html.
[3] See Unidata at http://www.unidata.ucar.edu/software/netcdf/.

9.1.2 Challenges

To develop the system, the following challenges should be addressed.

- *Building a scalable and elastic data warehouse to facilitate the management of massive datasets*—The data warehouse should offer two fundamental functions. First, it should be able to archive large volume climate datasets. Using ModelE as an example, the data volume of monthly outputs of ten-year simulations from 300 model runs is about 2.5 terabytes (Sun et al. 2012). Second, the configurations of the data warehouse should be dynamically adjusted according to the patterns of data operations such as uploading, deleting, and downloading data. The size of the warehouse should increase when most runs are toward completion to archive the income model outputs uploaded from citizens' computing devices. Therefore, a scalable and elastic data warehouse should be built to manage the big climate data.
- *Handling distributed data communications and concurrent access to support online geovisual analytics*—The Climate@Home system provides an online geovisual analytical module. While several geovisual analytical tools have been developed for scientists to conduct statistical analysis on large-volume climate data (e.g., Santos et al. 2013), these tools do not provide sufficient functions for interactive analysis on large volume spatiotemporal climate to datasets in a network-based environment. This is partially attributed to the fact that these tools were not designed to perform visualization and analysis in a Web environment. The capabilities of responding to concurrent requests and performing visualization tasks on massive datasets become the major bottlenecks of migrating traditional geovisual analytical functions into the network-based environment. Therefore, developing a mechanism that can handle distributed data communications and concurrent access is critical to the executing of the analytical pipeline.
- *Providing reliable and robust centralized management cyberinfrastructure for resources*—Centralized management of resources ensures delivering these resources dynamically within a shared network. Effective centralized management enhances the overall efficiency of the system. This project includes multiple types of resources such as data, model runs, and computing devices. The heterogeneity among those resources increases the difficulties of accomplishing a consolidated centralized management plan. It also raises challenges in the reliability and robustness of the cyberinfrastructure that supports centralized management.

To address the challenges discussed above, the spatial Web portal and the BOINC server are deployed on cloud services. The deployment improves the system performance and enhances the reliability of the Climate@Home system.

9.2 DEPLOYMENT AND OPTIMIZATIONS

This section describes the general workflow of deploying the two components onto Amazon EC2. The special configurations of the deployment procedure are highlighted.

9.2.1 General deployment workflow

9.2.1.1 Deploying the spatial Web portal

Amazon EC2 is used as the cloud service for the deployment (Figure 9.3). An AWS account is required before the deployment. The steps for creating an account are described in Chapter 4. Once an account is created, consumers first log in to the Amazon Web console and then start the deployment. The following steps demonstrate the general procedure for deploying the spatial Web portal, please refer to the Chapter 9 online video for detailed steps and commands.

- *Step 1. Authorize network access*—As described in Chapter 4, Section 4.3.1, this step authorizes the required network access by opening appropriate ports. Three ports should be opened: Port 22, Port 80, and Port 8080. Port 22 is used to enable access to the instance through the Secure Shell (SSH). Port 80 is required by the Apache2 Web server, Port 8080 is used by the Tomcat server.

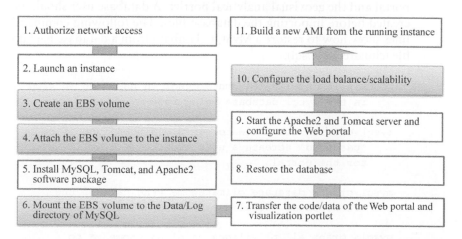

Figure 9.3 The process of deploying Climate@Home spatial Web portal onto Amazon EC2. (Gray boxes indicate the optional steps, which are described in Chapter 8.)

- *Step 2. Launch an instance*—A public Amazon Machine Image (AMI) with Ubuntu 12.04 as the OS is selected as the base image to create an instance for further customization. An instance is launched based on the public AMI.
- *Steps 3, 4, and 6. Configure Elastic Block Storage (EBS)*—Steps 3, 4, and 6 are optional. The detailed configurations have been discussed in Chapter 8, Section 8.2.2.1.
- *Step 5. Install software packages*—Several software packages are installed and configured as required by the spatial Web portal, including the Apache2 Web server, MySQL, and the Tomcat server. Installation of the Apache2 Web server and MySQL can be found in Chapter 5, Section 5.3.1. Installation of the Tomcat server can be found in Chapter 8, Section 8.2.
- *Step 7. Transfer the source codes and the database of the Climate@ Home system*—The Drupal source codes are moved to the Web folder of the Apache2 Web server. The geovisual analytical portlets are moved to the Web folder of the Tomcat server. Database files should be moved to a secured folder in the instance for later restoration. The database is dumped from the existing Web site. Transferring data and files from a local desktop to a cloud server can be found in Chapter 4, Section 4.3.

```
root@ip-10-189-149-104:/mnt$ mysqldump -u username
    -p climatehome > climatehome .sql # dump the database
    "climatehome" using the root account
```

- *Step 8. Restore the database*—To restore the database, database files are needed. This step restores the databases for both the main portal and the geovisual analytical portlet. A database user should be created before importing the database files. The following command lines show how to restore a MySQL database from a database backup file (climatehome.sql).

```
root@ip-10-189-149-104:/mnt$ mysql -u username -p #log
    in the mysql database using root

mysql>create user 'climateuser'@localhost identified by
    'password' #create a user account (e.g. climateuser)
    for the portal

mysql>create database climatehome #the name of the
    climate@home drupal database

mysql> grant all privileges on climatehome .* to
    'climateuser'@localhost identified by 'password'
    #grant permission to the database
```

```
mysql>flush privileges

mysql>exit #quit mysql

root@ip-10-189-149-104:/mnt$ mysql -u username -p cli-
    matehome< climatehome.sql #import the climatehome
    database climatehome.sql on the cloud server
```

- *Step 9.* Start the Apache2 and Tomcat server and configure the Web portal. The following commands can be used to start the servers on the Ubuntu instance. The first line starts the Apache2 Web server. To start the Tomcat server, one approach is to go to the bin directory of Tomcat and run the second command line.

```
root@ip-10-189-149-104:/$ sudo service apache2 start
root@ip-10-189-149-104:/Tomcat/bin/$ ./startup.sh
```

Once the two servers are started, the Web portal can be accessed from http://Your_VM_Public_DNS(IP)/climate@home/. The visualization portlet needs to be configured in the Web portal for proper display. The detailed steps for the configuration and the log in information for the Web portal can be found from the online material of Chapter 9.

- *Step 10. Configure the load balance or scalability*—This is an optional step to make the system more flexible and scalable. The configuration is introduced in Section 8.2.2.2 and 8.2.2.3.
- *Step 11. Create an AMI from the running instance*—The last step of the deployment is to create a new AMI based on the running instance of the Climate@Home portal. Section 4.3.1 (Chapter 4) describes the creation of an AMI. The image can be reused to create multiple virtual instances.

9.2.1.2 Deploying the BOINC server

The major steps of deploying the Climate@Home BOINC server are outlined in Figure 9.4. Please refer to the Chapter 9 online video for detailed steps and commands used in the steps below.

- *Step 1. Authorize network access*—Two ports should be opened: Port 22 and Port 80. Port 22 is used to access the instance through the Secure Shell (SSH). Port 80 is for HTTP.
- *Step 2. Launch an instance*—A public AMI with Ubuntu 12.04 as the OS is selected as the base image for creating an instance for further customization. Because the BOINC server manages intensive and high concurrent data transmissions between the client and

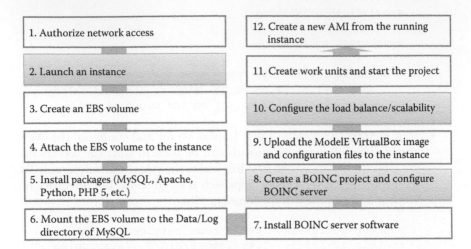

Figure 9.4 The process of deploying the BOINC server onto Amazon EC2. (Gray boxes indicate the special considerations described in Section 9.3.2.)

the server, the hardware configuration of the instance is critical. How to select an AMI for a BOINC server instance will be discussed in Section 9.2.2.

- *Steps 3, 4, and 6. Configure EBS*—The running of the Climate@ Home BOINC server uses the distributed FTPS server to store the uploaded results. Hence, EBS is optional for the BOINC server.
- *Step 5. Install software packages*—Several software packages are needed for the BOINC to function, including MySQL, the Apache2 Web server, Python, php5, and others.[1] Installation of the Apache2 Web server and MySQL can be found in Chapter 5, Section 5.3.1. For the installation of the other packages, please refer to the online content of this book.
- *Step 7. Install BOINC server software*—Please refer to the online content of this book for detailed instruction.
- *Step 8. Create a BOINC project and configure the BOINC server*— Upon finishing the installation of the BOINC server, consumers should create a BOINC test project, which distributes the ModelE applications to computing devices contributed by citizens.
- *Step 9. Upload the ModelE VirtualBox image and ModelE configuration files to the instance*—Two types of files should be uploaded to the server. The ModelE VirtualBox Image is put into the application folders[2] of the BOINC server. The ModelE configuration files are put into the

---

[1] See BOINC Software Prerequisites at http://boinc.berkeley.edu/trac/wiki/Software PrereqsUnix.

[2] See BOINC Platforms at http://boinc.berkeley.edu/trac/wiki/BoincPlatforms.

download folder of the "climateathome" project created in Step 8. See Chapter 4, Section 4.3.1 for information regarding uploading the files.

- *Step 10. Configure the load balance or scalability*—This is an optional step to make the BOINC server more flexible and scalable. The configuration is introduced in Section 8.2.2.2 and Section 8.2.2.3.
- *Step 11. Create work units and start the project*—Consumers need to create input and output templates to start the project.[1] Appendix 9.1 shows an example of the input template (E4M20a_000040_wu.xml) and Appendix 9.2 shows the output template (climateathome_re.xml). Using the two templates and the files in Step 9, consumers can create a work unit and start the project as shown below.

```
## Create work unit and start project

boincadm@ip-10-189-149-104:~/projects/climateathome $
    bin/create_work -appname modelE -wu_name
    E4M20a_00040e -wu_template templates/
    E4M20a_000040_wu.xml -result_template templates/
    climateathome_re.xml E4M20a_000040.R

boincadm@ip-10-189-149-104:~/projects/climateathome $
    bin/start
```

- *Step 12. Create an AMI from the running instance*—This step creates a new AMI based on the running instance of the BOINC server for reuses. Please see Chapter 4, Section 4.3 for creating an AMI.

9.2.2 Special considerations

According to the deployment procedures, network communications and robustness of the virtual instances are critical in ensuring the success of the system. When deploying the spatial Web portal and the BOINC server on cloud services, special considerations should be given to the following aspects:

- *Portal configuration*—The spatial Web portal can be hosted by different virtual machine instances. To minimize the impact of switching hosting servers, two approaches are recommended: (1) use relative paths instead of absolute paths and (2) use "localhost" as the account for database user. Both approaches ensure secured access and resolve the conflicts when changing hosts.
- *Backup data and VMs*—The data files of MySQL are kept on a separate EBS volume to provide persistent storage in the event of instance failure. Section 8.2 describes the process of creating and attaching

---

[1] See BOINC Input and Output Templates at http://boinc.berkeley.edu/trac/wiki/JobTemplates.

an EBS volume to the Climate@Home portal instance. Once an EBS volume is mounted to the data and log directory of MySQL, it can be cloned and used to restore the database on a second EC2 instance. The EBS volume can help restore another Climate@Home portal in case the current running portal on the cloud crashed. The utilization of the on-demand storage capabilities of the cloud is described in detail in Chapter 8, Section 8.2.

- *Load balance and auto-scaling*—To handle the massive requests, the Climate@Home system can take advantage of on-demand cloud computing capability. Load balance and auto-scaling should be considered. Chapter 8, Section 8.2, provides the detailed steps to set up the load balance and auto-scaling for cloud services.
- *AMI selection for the BOINC server*—Several hardware configurations of the virtual instance should be considered: CPU, memory, and storage. Within the framework of the Climate@Home system, the BOINC server mainly acts as a task scheduler and the model runs are distributed and executed at the client side. An instance with 8 gigabytes (GB) RAM and a 2-core CPU is sufficient. However, the data storage should be large enough to achieve terabytes to petabytes of climate datasets that are uploaded from the client side. The Amazon EBS, which provides highly available, reliable, and predictable storage volumes should be used in the EC2 instance. In addition, due to the underlying massive data transfer to the BOINC server, an instance with high Internet I/O performance and high storage I/O performance is preferred.

9.2.3 The differences from the general steps in Chapter 5

Compared to the general steps discussed in Chapter 5, the entire deployment procedure of the Climate@Home system to a cloud service has the following characteristics:

- *Deploying multiple components.* This system includes a grid engine server (BOINC server) and a spatial Web portal. The portal integrates several complex components such as the geovisual analytical portlet, interfaces to database management, and a Drupal-based CMS. Setting up a cloud-based server to host the portal involves configurations of multiple types of software packages.
- *Deploying and configuring multiple servers.* Within the spatial Web portal, two different servlet containers, which are the Apache2 Web server and the Tomcat server, are utilized to host two different components. These two containers may cause conflicts such as port arrangements.

- *Handling computing and network intensity.* Handling interactive geovisual analytics on massive data is both communication intensive and computing intensive. The deployment should consider the scalability and load balance in response to massive concurrent access. In addition, the efficiency of processing requests should be considered when selecting the AMI.

9.3 SYSTEM DEMONSTRATIONS

This section describes the functions of three major components of the system. They are the spatial Web portal, the geovisual analytical portlet, and the visual-based resource management portlet.

9.3.1 Overview of the spatial Web portal

The spatial Web portal provides a project entry point to support the interaction between the system and end users (Figure 9.5). It offers functions such as registering user accounts, downloading the BOINC executable, and forums for communications. Social media tools such as Facebook and Twitter are integrated to link with the portal to share information with the public for education and outreach. The usage of the two major portlets, the

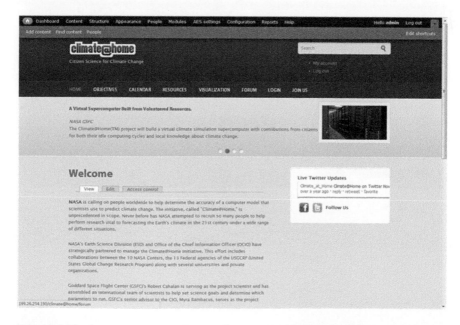

Figure 9.5 The home page of the Climate@Home spatial Web portal.

geovisual analytical portlet, and the resource management portlet, will be discussed in the next two subsections.

9.3.2 The geovisual analytical portlet

The geovisual analytical portlet provides a series of visual analytical functions. Multidimensional visualization tools, including NASA World Wind, Microsoft Bing Maps, Google Maps, and Google Earth, are integrated to facilitate visual explorations of NetCDF data. The netCDF Operators (NCO) are employed to support the statistical analyses functions, such as mean value calculations, anomaly detections, and simulation quality evaluations. Figure 9.6 shows the ten-year seasonal mean (summer) of precipitation for two selected areas of interest from 1951 to 1960 (Li et al. 2013).

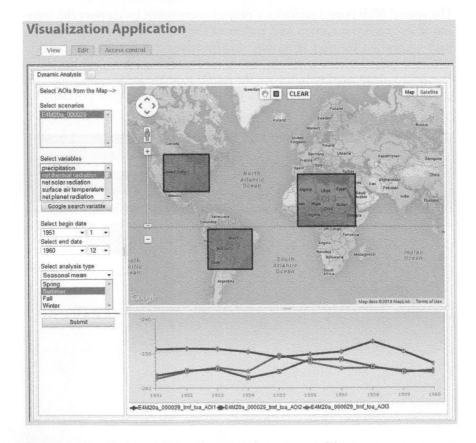

Figure 9.6 (See color insert.) Seasonal mean analysis using model outputs.

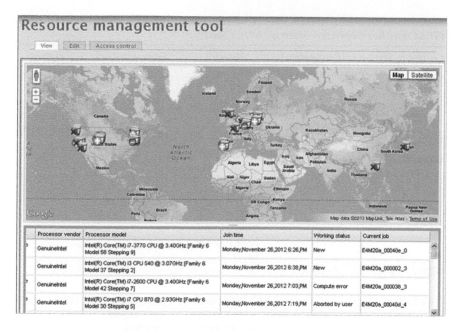

Figure 9.7 (See color insert.) Visualizing volunteered computing nodes with Google Maps.

9.3.3 The resource management portlet

Similar to the geovisual analytical portlet, the resource management port-let is designed to monitor, manage, and present the real-time information related to model running in a visual manner. Typical types of information are ModelE input configurations, information of users who contribute computing resources, the status of model running on each client-computing device, and an overall summary of model experiments. User information is stored in the spatial Web portal database when a user registers. Other computing related information is collected from the BOINC server database. Figure 9.7 illustrates the distribution of volunteered computing nodes across the world as well as each node's configuration and current working status.

9.4 CONCLUSION

The Climate@Home project provides a systemic computing and data management solution to facilitate large-scale, long-term climate modeling and analysis. As two major components of the system, the spatial Web portal and the BOINC server have been deployed to cloud services. Table 9.1 summarizes the cloud services used in the deployment. Cloud computing

Table 9.1 An Overview of the Cloud Services in the Climate@Home System (The Costs Are Estimated Based on AWS Cloud[*])

Computing Services	Configuration	Network	Costs with AWS	Purpose
Cloud data storage	10 TB	N/A	$0.080/ GB-month (Amazon S3); $0.095/ GB-month (Amazon EBS)	Archive the model outputs and visualization results
Cloud Web server	8-core 2.33GHz;			
16G RAM	1G	$0.299/hour (rent 1 year) $0.236 (rent; 3 years); $0.5/ hour (hourly base)		Host the spatial Web portal
BOINC server	8-core 2.33GHz;			
16G RAM	1G	$0.299/hour (rent 1 year); $0.236 (rent 3 years); $0.5/ hour (hourly base)		Serve as the grid engine machine

[*]See Cloud Computing Adoption Advisory Tool at http://swp.gmu.edu:8080/ESIPCostModelProject/.

techniques augment the capabilities of the system through providing reliable, scalable, and cost-efficient cyberinfrastructure (Yang et al. 2011a).

- Using a cloud storage service to archive datasets from model simulations and visual analytics addresses the scalability of managing "big data." This is a cost-efficient way to archive massive data with fluctuate data volume (Zeng et al. 2009). The costs of storing the model outputs differ with data size, duration of usages, and storage types (Chapter 6, Section 6.2.2). For example, Amazon Simple Storage Service (S3) usually charges $0.095/GB-month (per GB for one month) for the first 1 TB datasets and charges $0.08/GB-month for the next 49 TB datasets. The cost for Amazon EBS is relatively higher compared to S3 because it can be easily attached to or detached from an EC2 instance. However, the volume of cloud data storage can be adjusted based on the volume of the model outputs.
- Deploying the spatial Web portal to the cloud service handles concurrent access to the online geovisual analytical portlet. Through offering load balance and scalability, a cloud-based server can respond to massive requests efficiently (Chapter 17, Section 17.3). Multisite deployment

can even further contribute to the distributed concurrent access by providing high-speed network connection at a global scale. In addition, based on the cost analysis in Chapter 8, Section 8.4, the cloud host of a Web application would save the budget around 57% each year.

- Deploying the BOINC server in a cloud environment ensures the robustness, high performance, security, and reliability of managing model simulations. The BOINC server is responsible for managing computing resources and dispatching model simulation tasks on a routine basis. The reliability and robustness of the BOINC server are crucial in conducting huge amounts of long-term model runs. Most cloud services guarantees a high degree of availability. For example, Amazon EC2 Service Level Agreement (SLA) guarantees 99.95% availability for all Amazon cloud data centers (Chapter 11, Section 11.2.2).

There are a few issues pertaining to the utilization of computing techniques that are worth further investigation:

- Develop a user evaluation component in the system that allows users, including both citizen contributors and scientists, to test the functions and the performance of the system. Based on the statistics collected from user experience assessments, researchers can better examine the system and redesign existing components.
- Develop an efficient computing framework for utilizing the heterogeneous computing resources contributed by citizens. In addition to the availability of computing devices contributed from citizens, the grid engine should dispatch tasks based on the reliability, the historical performance, and the network connectivity of these computing devices.
- Design and implement a cloud-based distributed data management strategy. In this project, the cloud storage has been merely used as the data warehouse to store datasets. The scalability and elasticity of cloud storage services have not been fully explored yet.
- Explore optimal deployment solutions of the spatial Web portal based on spatiotemporal principles. All deployment scenarios discussed in the chapter are relatively standard procedures without considering spatiotemporal principles such as patterns of user access. These principles are also critical factors that impact the performance of the system (Yang et al. 2011b).

9.5 SUMMARY

This chapter describes the development of a comprehensive system to facilitate climate modeling, with an emphasis on how cloud services can be used to enhance the capabilities of the system. Section 9.1 introduces the background, challenges, and components of the Climate@Home system. Section 9.2 describes the procedures of deploying the system onto cloud services with an emphasis on special procedures and considerations. Demonstrations of selected components are discussed in Section 9.3. Section 9.4 summarizes the benefits of utilizing cloud services to support the system as well as future research needs.

9.6 PROBLEMS

1. Using the Climate@Home system as an example, list the major system components required for supporting climate modeling and analysis in a distributed environment.
2. Using the development of the Climate@Home system as an example, explain why cloud computing techniques are essential in supporting climate modeling and analysis in a distributed environment.
3. What are the major steps of deploying the spatial Web portal onto the cloud services? In the process of deploying the spatial Web portal, what are the special considerations related to network configurations?
4. What are the major steps of deploying the BOINC server onto the cloud services? What are the special considerations compared to the general application deployment in Chapter 5?
5. What are the hardware configuration requirements of a virtual machine serving as a BOINC server?
6. What are the uses of an Elastic Block Store (EBS) in the cloud-enabled Climate@Home system?
7. List the cloud services used in the deployment of the Climate@Home system.
8. What are the remaining issues for deploying the Climate@Home components onto the cloud?

APPENDIX 9.1 E4M20A_000040_WU.XML

```
<!-- workunit template for the BOINC E4M20a_000040
     program -->
<file_info>
 <number>0</number>
</file_info>
<file_info>
 <number>1</number>
</file_info>
<workunit>
```

```
   <file_ref>
    <file_number>0</file_number>
    <open_name>shared/boinc_app</open_name>
    <copy_file>1</copy_file>
   </file_ref>
   <file_ref>
    <file_number>1</file_number>
    <open_name>shared/E4M20a_000040.R</open_name>
    <copy_file>1</copy_file>
   </file_ref>
   <min_quorum> 1 </min_quorum>
   <rsc_disk_bound> 10000000000 </rsc_disk_bound>
   <rsc_fpops_est>1444000000000000</rsc_fpops_est>
   <rsc_fpops_bound>1888000000000000</rsc_fpops_bound>
   <credit>3072</credit>
  </workunit>
```

APPENDIX 9.2 CLIMATEATHOME_RE.XML

```
    <file_info>
      <name><OUTFILE_0/></name>
      <generated_locally/>
      <upload_when_present/>
      <max_nbytes>8000000000</max_nbytes>
      <url><UPLOAD_URL/></url>
    </file_info>

<result>
    <file_ref>
        <file_name><OUTFILE_0/></file_name>
        <open_name>shared/out.tar.gz</open_name>
        <optional>1</optional>
        <no_validate>1</no_validate>
        <copy_file/>
    </file_ref>
</result>
```

REFERENCES

Anderson, D. P. 2004. BOINC: A system for public-resource computing and storage. In *Grid Computing. Proceedings of 5th IEEE/ACM International Workshop*, pp. 4–10.

Anderson, D. P. 2010. Volunteer Computing with BOINC. In Cerin, C., and Fedak, G., eds. *Desktop Grid Computing*. Boca Raton, FL: CRC Press/Chapman & Hall, pp. 3–25.

Li, Z., C. Yang, M. Sun et al. 2013. A High Performance Web-Based System for Analyzing and Visualizing Spatiotemporal Data for Climate Studies. In S. Liang, X. Wang, C. Claramunt, eds. *Web and Wireless Geographical Information Systems*, pp. 190–198. Heidelberg: Springer Berlin.

Santos, E., J. Poco, Y. Wei et al. 2013. UV-CDAT: Analyzing climate datasets from a user's perspective. *Computing in Science & Engineering* 15, no. 1: 94–103.

Schmidt, G. A., R. Ruedy, J. E. Hansen et al. 2006. Present-day atmospheric simulations using GISS ModelE: Comparison to *in situ*, satellite, and reanalysis data. *Journal of Climate* 19, no. 2: 153–192.

Skamarock, W. C. and J. B. Klemp. 2008. A time-split nonhydrostatic atmospheric model for weather research and forecasting applications. *Journal of Computational Physics* 227, no. 7: 3465–3485.

Sun, M., J. Li, C. Yang et al. 2012. A Web-based geovisual analytical system for climate studies. *Future Internet* 4, no. 4: 1069–1085.

Taylor, K. E. 2001. Summarizing multiple aspects of model performance in a single diagram. *Journal of Geophysical Research: Atmospheres (1984–2012)* 106, no. D7: 7183–7192.

Yang P., Evans J., Cole M., Alameh N., Marley S., and Bambacus M. 2007. The emerging concepts and applications of the spatial Web portal. *Photogrammetry & Remote Sensing* 73, no. 6: 691–698.

Yang, C., M. Goodchild, Q. Huang et al. 2011a. Spatial cloud computing: How can the geospatial sciences use and help shape cloud computing? *IJDE* 4, no. 4: 305–329.

Yang C., H. Wu, Z. Li, Q. Huang, and J. Li. 2011b. Utilizing spatial principles to optimize distributed computing for enabling physical science discoveries. *Proceedings of National Academy of Sciences* 108, no. 14: 5498–5503.

Zeng, W., Y. Zhao, K. Ou et al. 2009. Research on cloud storage architecture and key technologies. In *Proceedings of the 2nd International Conference on Interaction Sciences: Information Technology, Culture and Human*, pp. 1044–1048. ACM.

Chapter 10

Cloud-enabling dust storm forecasting

Qunying Huang, Jizhe Xia, Manzhu Yu,
Karl Benedict, and Myra Bambacus

Real-time dust storm forecasting is a typical application of computing intensities, which requires fast, high-resolution, and large geographic area simulation. This chapter introduces how to use cloud services to support dust storm simulation and predictions.

10.1 DUST STORM MODELING: BACKGROUND AND CHALLENGES

10.1.1 Background

Dust storms result from strong turbulent wind systems entraining dust particles into the air and reducing visibility from miles down to several meters (Goudie and Middleton 1992). Global climate change has driven up the frequency and intensity of dust storms in the past decades with negative consequences on the environment, human health, and physical assets. For example, severe dust storms impact the southwestern United States several times per year and cause major vehicle accidents, property damage, injuries, and loss of lives (Shoemaker and Davis 2008). The negative impacts of dust storms have motivated scientists to develop models to better understand and predict the distribution and intensity of dust emission, deposition, and structure. Since the late 1980s, several research groups have developed dust models that can correctly predict spatiotemporal patterns, evolution, and the magnitude order of dust concentration, emissions, and deposition (Huang et al. 2013a). In addition to weather and climate prediction, dust storm modeling and forecasting could also be used in practice (WMO 2012) to: (1) provide important implications of air quality for constituencies ranging from the public to industry; (2) assist public health services with improved operational disease surveillance and early detection; (3) improve air and highway safety and accident emergency management; and (4) serve as a warning advisory and assessment system to give instructions for adaptive action, such as changing the time of planning, strengthening infrastructure, and construction of windbreaks and shelterbelts.

10.1.2 Challenges

Utilizing existing dust models to predict high-resolution dust storms for large geographic areas poses several critical challenges:

- Simulating dust storm phenomena is complex and computing intensive (Xie et al. 2010). The periodic phenomena simulation requires repeating the same set of numerical calculations with a time interval for many times. Figure 10.1 shows a typical procedure for a Nonhydrostatic Mesoscale Model (NMM)-dust model (Huang et al. 2013a), which is a popular dust storm forecasting model. Each iteration includes 21 module calculations. For a given domain size, the computing complexity of an atmospheric model behaves as a function of where n is the grid dimension, including one time dimension, two horizontal dimensions, and one vertical dimension (Baillie et al. 1995). If the time dimension is extended, more iterations are needed. Doubling the geographic area on the horizontal direction would result in a fourfold increase in computing cost. Doubling the spatial resolution (e.g., from 10 km to 5 km) would result in an eightfold increase in computational cost because it would also require doubling the time step to keep the model accuracy (Baillie et al. 1995). High performance computing is used to perform dust storm simulation for large geographical areas with high resolutions (Huang et al. 2013a). NMM-dust (Huang et al. 2013a) is parallelized to leverage high performance computing (HPC) by decomposing the domain into multiple subdomains and distributing the computing load of each subdomain onto one CPU core as a process. The core that processes a subdomain must communicate with the neighboring cores for synchronization. Subroutines shown in Figure 10.1 in gray boxes require intensive data exchange and synchronization if parallelized (Huang et al. 2013a).

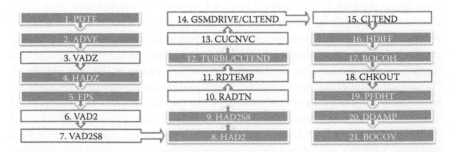

Figure 10.1 Computing subroutines for the NMM-dust model. (The subroutines in gray boxes require communication and synchronization.) (From Huang et al. 2013a.)

- Dust storm forecasting is a time critical task that needs to be finished in a limited time frame. For example, a two-hour computing limit is recommended for one-day forecasting to make the results useful (Drake and Foster 1995). Limited geographic area and/or resolution forecasting is usually performed to complete the simulations within the time limit (Wolters, Cats, and Gustafsson 1995). However, a zip code level resolution is needed for dust storm forecasting to support emergency decision making for governmental agencies, such as preparing medications for public health (Xie et al. 2010). Figure 10.2 shows the computing time required for a 24-hour forecasting over different domain sizes on the horizontal directions and with the same vertical layers (37 layers) and spatial resolution (3 km). More than 12.6 hours are needed to forecast a 10 × 10 degree domain size using an HPC cluster with 25 computing nodes (Yang et al. 2011). It is estimated that forecasting the whole southwest United States with a domain size of 37 × 20 degrees would require about 93 hours. Such computing performance is not acceptable because we would be forecasting the dust event in the past.
- Dust storms are disruptive events. It is estimated that the total time of dust storms in one year was generally less than 90 hours and only takes less than 1% of one year assuming that each dust storm lasts on average two hours (NOAA 2011). Therefore, a forecasting system for such events would expect different computing and access requirements during different times of the year and even different hours within a day.

To tackle these challenges, cloud computing can be leveraged. Cloud services can handle the spike and disruptive computing requirements with elastic and on-demand computing resources (Huang et al. 2013b). Additionally, with the development of cloud computing technologies such as virtualization and the general advancement of cloud infrastructure, cloud services are ready to support computing intensive applications.

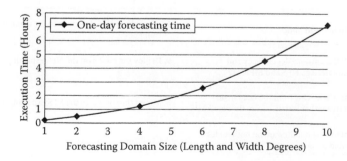

Figure 10.2 Execution time for different geographic domain forecasts. (From Huang et al. 2013b.)

For example, Amazon EC2 offers cluster instances with a 10 Gbps network connection. Each cluster instance has a minimum of 23 GB RAM, and two quad-core processors with a clock speed of 2.93 GHz. Such hardware and network configurations are far better than grid computing environments with heterogeneous computing resources and a Wide Area Network (WAN) connection, and even better than most private homogenous HPC cluster configurations. This makes cloud computing a new advantageous computing paradigm to resolve scientific problems that are computing intensive and traditionally require a special high-performance cluster (Rehr et al. 2010).

10.2 DEPLOYMENT AND OPTIMIZATION

10.2.1 General workflow

The general steps to deploy a dust storm model onto Amazon EC2 are as follows (Figure 10.3):

> *Step 1. Authorize network access*—Network access authorization was introduced in Chapter 4. Port 22 is opened to enable communication between the local server and EC2 instances through Secure Shell (SSH).
>
> *Step 2. Launch an instance*—Launch an instance as the head node of the dust storm simulation task (see Chapter 4, Section 4.3 for reference). A cluster instance with eight CPU cores and 23 GB of memory is chosen.

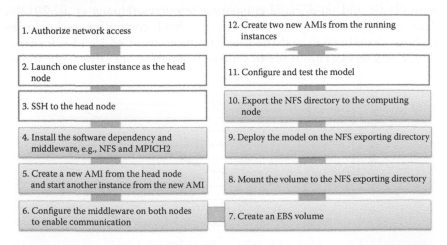

Figure 10.3 The process of deploying the dust storm model onto Amazon EC2. (Gray boxes indicate that the step requires special consideration.)

Step 3. SSH to the instance. Use an SSH tool to access the instance created in Step 2. Instance access was discussed in detail in Chapter 4, Section 4.3.

Step 4. Install software dependencies—Dust storm model execution requires that the Network File System (NFS) and MPICH2[1] are pre-installed on the head node and computing nodes. NFS enables all other computing instances to share the software package and model data. By using NFS, we only need to set up the model on the head node while other computing instances can share this environment without installing and configuring a model run environment. MPICH2 is one of the most popular implementations of Message Passing Interface (MPI). The following command line introduces the process of installing NFS and MPICH2 in an Amazon EC2 instance with CentOS 5.6 as the operation system.

```
[root@domU-head~] yum install gcc gcc-c++ autoconf
    automake
[root@domU-head~] yum -y install nfs-utilsnfs-utils-
    lib sytem-config-nfs #install NFS
[root@domU-head~] tar -zvxf mpich2-1.5.tar.gz
    download MPICH2 package and unzip
[root@domU-head~] mkdir /home/clouduser/mpich2-
    install # create an installation directory
[root@domU-head~] cd mpich2-1.5.tar.gz
[root@domU-head~] ./configure -prefix=/home/
    clouduser/mpich2-install
[root@domU-head~] make # Build MPICH2
[root@domU-head~] make install # Install MPICH2
```

After NFS is installed, the following commands can be used to configure and start NFS.

```
[root@domU-head ~] mkidr /headMnt # create a NFS
    export directory
[root@domU-head ~] echo "/headMnt *(rw) " >> /etc/
    exports
[root@domU-head ~] exportfs -ra
[root@domU-head ~] service nfs start #start up NFS
```

Step 5. Create an AMI and a computing node—Build an AMI based on the configured instance in Step 4. Use this new AMI to launch one instance as the computing node.

Step 6. Configure the head node and computing node—This step establishes network communication between the head node and computing nodes. First, use SSH to access the head node

---

[1] See MPICH2 at http://www.mcs.anl.gov/research/projects/mpich2staging/goodell/index.php.

and add the IP of the computing node to the host list. Then create a public key at the head node. Finally, use SSH to access each computing node, and copy the head node's public key to the authorized key list.

```
[root@domU-headMnt ~] vi /etc/hosts #access to the
    hosts list of the head node
[root@domU-headMnt ~] ssh-keygen -t rsa #create a
    public key at the head node

[root@domU-computing~] mkdir -p /root/.ssh/
[root@domU-computing ~] scp root@domU-head: /root/.
    ssh/id_ras.pub /root/.ssh/ #copy the public key
    from the head node to the computing node
[root@domU-computing ~] cat /root/.ssh/id_ras.pub >>
    /root/.ssh/authorized_keys
```

Step 7. Create an EBS volume and attach it to the head node—Please refer to Chapter 8, Section 8.3 for information regarding creating a new EBS volume and attaching it to a running instance.

Step 8. Mount the volume to the NFS export directory—Please refer to Chapter 8, Section 8.3 for mounting the volume to an instance.

Step 9. Deploy the model on the NFS export directory—Transfer the dust storm model (refer to Chapter 5, Section 5.3.1 for large data transfer between the local server and EC2 instances) to the NFS export directory of the head node, which is /headMnt in this case.

Step 10. Export the NFS directory to the computing node—This step establishes the NFS connection between the head node and computing node, which enables the computing node to share the package suites from the head node. Use SSH to access the computing node, and mount the volume to the NFS directory. The command below shows an example of mounting a computing node volume to the NFS directory. domU-head is the domain name of the head node. /headMnt is the file path of the NFS directory in the head and computing nodes. After issuing the command to mount the NFS directory from the head node to the computing node, cloud consumers can go to the /headMnt directory and check if the model is available on the computing node.

```
[root@domU-computing ~] mkdir //headMnt # create the
    directory
[root@domU-computing ~] mount -t nfs -o rw domU-head:/
    headMnt //headMnt #Mount the volume to the NFS export
    directory
```

Step 11. Configure and test the model—Create a host file with a list of either IPs or domain names that can run MPI tasks (e.g., mpd. hosts), and add the private IP of each computing node to this file as shown below (including two computing nodes).

```
199.26.254.161 # node1
199.26.254.162 # node2
```

On the head node, start the MPI connection. The below command shows how to start the MPICH2 service. The parameter -n 3 is the number of computing nodes plus head node, and mpd.hosts is the machine configuration file. The below command also shows an example of executing the model script: run_test.sh is the model execution script, and out.log is the log file to store model execution information.

```
[root@domU-head~] mpdboot -n 3 -f mpd.hosts -v #
    start up MPICH2
[root@domU-xxx~] ./run_test.sh >& out.log & # Run the
    model
```

Step 12. Create an AMI from the running instance—Finally, two new AMIs can be created based on the running head node and computing node separately (refer to Chapter 4, Section 4.3 for how to create an AMI).

10.2.2 Special considerations

In addition to the general deployment steps, several configurations are unique to dust storm models with high computing power requirements.

- *Configure a virtual cluster environment*—To support dust storm forecasting with cloud services, a virtual cluster environment should be set up. As introduced in Chapter 5, Section 5.3, one or more virtual instances should be used with middleware installed and configured to construct a virtual cluster. In this chapter, MPICH2 is used as the middleware (Chapter 5 uses Condor) to dispatch the model run tasks and collect the results. In order to save the time to configure the computing nodes, an image from the head node at Step 5 (Figure 10.3) can be built to launch a computing node. In addition, the communication overhead can be greatly reduced if all instances are launched within the same data centers and on physically close machines. Therefore, cloud consumers should create a placement group and launch multiple cluster instances into the placement group (Figure 10.4). A cluster placement group is a logical grouping of cluster

Figure 10.4 (See color insert.) Create a placement group and place the instances within the same group.

instances within a single EC2 data center. A placement group cannot span multiple data centers. In this way, consumers can logically group cluster instances into clusters in the same data center. It enables many HPC applications to get the full-bisection bandwidth and low-latency network performance required for tightly coupled, node-to-node communication. Amazon recommends that the cloud consumers launch the minimum number of instances that are required for supporting the applications in the placement group in a single launch request. If consumers launch only a few instances and try to add more instances to the placement group later, they may get an "Insufficient Capacity" error because no physical resources are available to create virtual machines on the same HPC rack. If consumers do receive an "Insufficient Capacity" error, stop and restart the instances in the placement group, and try to add more instances again.[1]

Cloud consumers can also use the following command line to create a placement group (i.e., VirtualCluster) and place all instances (e.g., two instances in this case in the same group).

```
PROMPT>ec2-create-placement-group VirtualCluster -s
cluster
PROMPT>ec2-run-instances ami-xxx -n 2 --placement-
group VirtualCluster
```

- *Loosely coupled nested modeling and cloud computing*—During the loosely coupled nested model test (Huang et al. 2013a), a large domain is reduced into multiple small regions, which are identified by the coarse model ETA-8bin. A large amount of computing resources should be

---

[1] See AWS Elastic Cloud Compute at http://docs.aws.amazon.com/AWSEC2/latest/UserGuide/using_cluster_computing.html.

leveraged to concurrently run these 18 areas of interest (AOIs), each requiring high-resolution forecasting. Elastic Amazon EC2 cluster instances are utilized to enable the nested modeling approach to its full extent. This is because multiple virtual clusters can be constructed in the Amazon EC2 platform with each virtual cluster running one subregion directly (Section 10.3.3 covers the performance details).

- *Auto-scaling*—Since this is not a Web application, the method to trigger auto-scaling resources through CloudFormation (Chapter 8, Section 8.3) is not suitable. The cloud consumer would have to write scripts with the EC2 APIs to launch more instances to handle the computing when needed.

10.2.3 Summary of the differences from the general steps in Chapter 5

Compared to the general steps discussed in Chapter 5, the entire deployment procedure of a dust storm model to a cloud service has the following characteristics:

- *Handling computing intensity*—This dust storm model requires leveraging multiple virtual instances to run the simulation concurrently. Therefore, a virtual cluster environment should be set up with the middleware MPICH2 installed and configured properly on the head node and computing nodes.
- *Handling communication intensity*—A placement group (Figure 10.4) should be created and all instances for the virtual cluster should be within the same group to reduce the communication overhead.

10.3 DEMONSTRATION

10.3.1 Phoenix dust storm event

The State of Arizona is home to some of the most spectacular displays of blowing dust in the United States. Dangerous dust storms impact the state several times per year and cause numerous economic losses (Shoemaker and Davis 2008). The largest of these dust storms is called a *haboob*, which is the most common in the central deserts of Arizona during its summer season, with the frequency of occurrence peaking in late July and early August. The city of Phoenix, Arizona, experiences on average about three haboobs per year during the months of June to September (Shoemaker and Davis 2008). Figure 10.5 shows a classic haboob that moved through the Phoenix area on July 1, 2007. This chapter uses this event as an example to show how a dust storm model can simulate a dust storm event and how cloud computing can help perform such a simulation.

Figure 10.5 (See color insert.) Photograph of a haboob that hit Phoenix, Arizona, on July 1, 2007. (Courtesy of Osha Gray Davidson, *Mother Jones*, San Francisco, CA at http://www.motherjones.com/blue-marble/2009/09/australias-climate-chaos.)

10.3.2 Simulation result

Figure 10.6 shows a time-frame result at 3 A.M., July 2, 2007, produced by ETA-8bin (a low-resolution model) (Huang et al. 2013a) and NMM-dust models in different spatial resolutions. The model results show similar patterns for the dust storm's area and NMM-dust results with a 3-km resolution, which picked up much more detailed information about the dust concentrations.

10.3.3 Performance

Elastic Amazon EC2 cluster instances are utilized to enable the loosely coupled nesting approach to its full extent (Huang et al. 2013b). During the loosely coupled nesting approach, ETA-8bin performs a quick forecasting with low resolution (22 km) to identify potential hotspots with high dust concentration. NMM-dust will perform high resolution (3 km) forecasting over the much smaller hotspots in parallel to reduce computational requirements and computing time.

In this case, 18 AOIs are identified by the coarse model ETA-8bin. Figure 10.7 shows the width and length distribution of those 18 AOIs. It is observed that most of them are distributed within a 2 × 2 degree scope. Therefore, 18 cluster instances are launched from the AMI configured through the steps in Section 10.2 and each instance is responsible for handling the simulation for one AOI.

Figure 10.8 shows the execution time for each instance by 18 Amazon instances, with each instance simulating one AOI region for a 24-hour forecast. The results reveal that most of the AOIs can be successfully

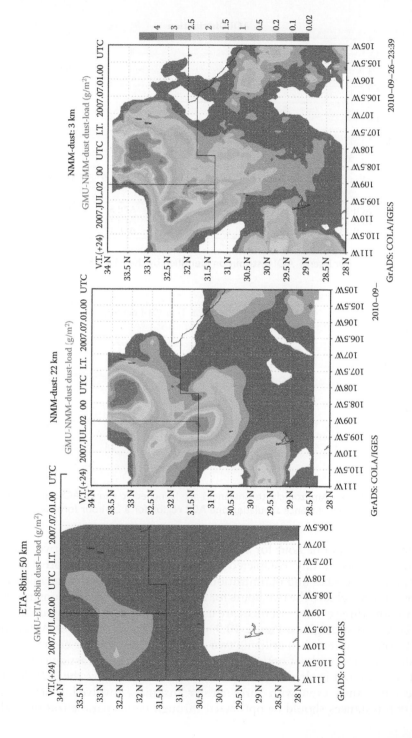

Figure 10.6 (See color insert.) Comparison of the simulation results by ETA-8bin and NMM-dust AOI 10, 11, 12, and 13 at 3 A.M., July 2, 2007.

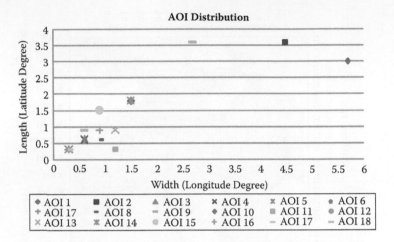

Figure 10.7 (See color insert.) AOI distribution identified by the ETA-8bin.

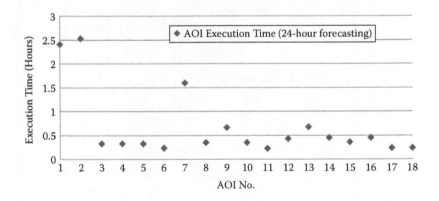

Figure 10.8 Eighteen Amazon cluster instances are launched to run 18 AOIs in parallel with each instance simulating one AOI region for a 24-hour forecast.

completed within one hour for a 24-hour forecasting. However, two AOIs cannot be successfully completed within two hours. Therefore, more computing resources and optimizations should be integrated to enable the computing to achieve the two hours for 24-hour simulation time constraint.

Two and three cluster instances are used to further test the computability of those two subregions. Figure 10.9 shows the performance of Amazon EC2 cloud services when utilizing two and three instances for each virtual cluster with and without optimizations for three-hour forecasting over the first subregion. It is observed that the system cannot achieve the 0.25-hour time constraints even with three cluster instances involved. Therefore, the two instances should be optimized through better parallelization and

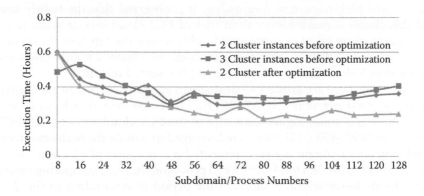

Figure 10.9 Performance comparisons of Amazon cluster instances before and after optimizations.

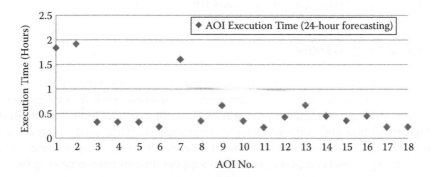

Figure 10.10 Eighteen subregions run on Amazon EC2 cloud service after optimization. Both AOI 1 and 2 utilize two optimized cluster instances.

scheduling strategies, including neighbor mapping and spatiotemporal optimization strategies (Yang et al. 2011). Figure 10.9 shows that the two optimized instances can successfully complete the forecasting within 0.25 hours if using more than 64 processes. Finally, with these two instances starting 64 processes, it is observed that the two AOIs can be successfully completed within two hours (Figure 10.10).

10.3.4 Cost-efficiency

For an operational dust storm forecasting system, usually an EC2 instance is reserved for long-term forecasting with a much lower price compared to an on-demand instance. It costs only $0.16 per hour for general public user access and low-resolution forecasting. When the dust storm events occur, a large group of instances would be started to respond to concurrent user access and high-resolution forecasting. It is observed that the hourly cost of this cluster is much higher than that of EC2 cloud service when there is no dust storm event occurring and only one reserved instance is needed (Huang et al. 2013b). Under this situation, the cluster's hourly costs are more than 896 times than that of the single reserved instance.

Typically, there are around 45 dust storm events occurring in the United States in total per year (NOAA 2011). The yearly cost of a local cluster is around 12.7 times higher than that of the EC2 cloud service if 28 EC2 instances (with 400 CPU cores) are leveraged to handle the high resolution and concurrent computing requirements for a duration of 48 hours. With much lower costs, the cloud service can provision the same computing power as a local cluster within a shorter time period in responding to the dust storm events. These cost comparison results indicate that it is economical to use cloud computing by maintaining low access and resolution forecasting while invoking a large scale of computing power to perform high-resolution forecasting for large geographic areas when needed.

10.4 CONCLUSION

Dust storm events exhibit an interannual variability and are typical disruptive events. The computing pool to host an operational dust storm forecasting system should also support the disruptiveness by (1) scaling up to respond to potential high resolution and concurrent computing requests when dust storm events happen, and (2) scaling down when no dust storm events occur to save costs and energy. Cloud computing is used to support dust storm forecasting by:

- Being capable of provisioning a large amount of computing power in a few minutes to satisfy the disruptive computing requirements of dust storm forecasting (Section 10.3.2).
- Economically sustaining low access rates and low-resolution models, while still being able to invoke a large amount of computing power to perform high-resolution forecasting for large public access when needed (Section 10.3.3).

More advanced configurations such as a better file system design to reduce I/O overhead (Huang et al. 2013a) can better support large-scale simulations such as dust storm forecasting. In the cluster computing architecture, each computing node is usually designed to access the same remote data storage to execute tasks in parallel. In this example, NFS is used to share the storage and to ensure the synchronization of data access. However, different file systems could have great impacts on the performance of I/O and data intensive applications. Therefore, cloud consumers can explore other solutions such as the Parallel Virtual File System version 2 (PVFS2) (Latham et al. 2010).

10.5 SUMMARY

This chapter discusses how cloud services are capable of supporting the forecasting of disruptive events such as dust storms, using Amazon EC2 as an example. Section 10.1 explains the background and motivation of conducting the dust storm forecasting, and presents the major challenges in performing dust storm forecasting. The details of the configuration and deployment of a dust storm model onto the EC2 cloud service are described in Section 10.2. Section 10.3 demonstrates how a cloud service can enable dust storm forecasting using the December 22, 2009 Phoenix dust storm as an example. Section 10.4 concludes the improvements by using cloud computing to support such large-scale simulation and forecasting as well as future research directions.

10.6 PROBLEMS

1. What are the computing challenges for dust storm forecasting?
2. What are the general steps to deploy dust storm simulation onto the cloud?
3. Which instance type is better for dust storm forecasting; regular instance or HPC instance? Why?
4. How to configure a virtual high performance computing (HPC) cluster to support computing-intensive applications?
5. How is Elastic Block Storage (EBS) service used in supporting the dust storm model deployment to the cloud?
6. How do you create a placement group for HPC instances using both Amazon Web Console management and command line tools? Why do we need this step?
7. Compared to the deployment workflow for general applications introduced in Chapter 5, what are the special considerations for a dust storm model?
8. Why can cloud computing achieve cost-efficiency?
9. Why does cloud computing provide a good solution to support a disruptive event (e.g., dust storm) simulation?

REFERENCES

Baillie, C. F., A. E. MacDonald, and S. Sun, 1995. QNH: A portable, massively parallel multi-scale meteorological model. *Proceedings of the 4th Int'l Conference on the Applications of High Performance Computers in Engineering*, June 19–21. Milan, Italy.

Goudie, A. S. and N. J. Middleton. 1992. The changing frequency of dust storms through time. *Climatic Change* 20, no. 3: 197–225.

Drake, J. and I. Foster. 1995. Introduction to the special issue on parallel computing in climate and weather modeling. *Parallel Computing* 21, no. 10: 1539–1544.

Huang, Q., C. Yang, K. Benedict, A. Rezgui, J. Xie, J. Xia, and S. Chen. 2013a. Using adaptively coupled models and high-performance computing for enabling the computability of dust storm forecasting. *International Journal of Geographic Information Science* 27, no. 4: 765–784.

Huang, Q., C. Yang, K. Benedict, S. Chen, A. Rezgui, and J. Xie. 2013b. Enabling dust storm forecasting using cloud computing. *International Journal of Digital Earth* 6, 4: 338–355.

Latham, R. et al. 2010. A next-generation parallel file system for Linux cluster. *Linux World Magazine* 2, no. 1: 54–57.

NOAA. 2011. Dust Storm Database [online]. http://www4.ncdc.noaa. gov/cgi-win/ wwcgi.dll?wwevent_storms (accessed October 30, 2011 and 2013).

Rehr, J. J., F. D. Vila, J. P. Gardner, L. Svec, and M. Prange. 2010. Scientific computing in the cloud. *Computing in Science & Engineering* 12, no. 3: 34–43.

Shoemaker, C. and J. Davis. 2008. Hazardous Weather Climatology for Arizona, NOAA Technical Memorandum. 2008. http://www.wrh.noaa.gov/wrh/tech-Memos/TM-282.pdf (accessed March 10, 2013).

WMO (World Meterological Organization). 2012. WMO Sand and Dust Storm Warning Advisory and Assessment System (SDS-WAS): Science and Implementation Plan 2011–2015. http://www.wmo.int/pages/prog/arep/ wwrp/new/documents/SDS_WAS_implementation_plan_01052012.pdf (accessed March 30, 2013).

Wolters, L., G. Cats, and N. Gustafsson. 1995. Data-parallel numerical weather forecasting. *Science Program* 4, no. 3: 141–153.

Xie, J., C. Yang, B. Zhou, and Q. Huang. 2010. High-performance computing for the simulation of dust storms. *Computers, Environment and Urban Systems* 34, no. 4: 278–290.

Yang, C., H. Wu, Q. Huang, Z. Li, and J. Li. 2011. Using spatial principles to optimize distributed computing for enabling the physical science discoveries. *Proceedings of the National Academy of Sciences* 108, no. 14: 5498–5503.

Part IV

Cloud computing status and readiness

This part uses five chapters to systematically examine the readiness of both commercial cloud services and open-source cloud computing solutions in supporting geoscience applications. Chapter 11 introduces three cloud services: Amazon EC2, Windows Azure, and NASA Nebula. Chapter 12 illustrates the methods and tools to test the readiness of the three cloud services. Chapter 13 covers four major open-source cloud computing solutions. Chapter 14 provides detailed coverage on methods and tools for benchmarking computing performance of the solutions. Finally, Chapter 15 reports the findings of GeoCloud, a cross-agency project led by Federal Geographic Data Committee (FGDC) to investigate how to deploy geoscience applications in the cloud, and the benefits and drawbacks of hosting applications in the cloud.

Chapter 11

Cloud services

Chen Xu, Jizhe Xia, Qunying Huang, and Myra Bambacus

This chapter introduces three Infrastructure as a Service (IaaS) cloud services: Amazon EC2, Windows Azure, and NASA Nebula. IaaS providers offer virtual machines (VMs) to cloud customers for hosting their data and applications. Several critical factors may impact the capacities of cloud providers to effectively offer services to their customers, such as geographic presence, user interface, automatic scaling and load balancing, and service level agreement (SLA).

11.1 INTRODUCTION TO CLOUD SERVICES

11.1.1 Geographic presence

Industrial companies are pioneers in service-based computing paradigm implementation. From the early time-sharing mode to the recent virtualization-based cloud computing mode, commercial companies have populated advanced computing-sharing technology. Although cloud service is still at the early stage of its life cycle, both cloud providers and cloud consumers are confident about the potential. For example, cloud computing is used to help organizations reach more distributed markets, reduce investment in hardware, and be more agile to gain a competitive advantage (ISACA 2012).

Virtualization enables the geographic distribution of computing facilities (Chapter 3) for cloud consumers to collocate their services with end users. Major cloud providers usually have service centers in multiple geographical locations. For example, Amazon and Microsoft have multiple data centers around the world. Cloud consumers therefore will have better service continuity, for example, by planning better computing and data storage services closer to end users, or by quickly shifting services to another center at a different geographical location when one data center is disconnected. Li et al. (2010) and Yang et al. (2011) demonstrate that the geographic presence of cloud providers becomes relevant to the selection of service providers because of the spatial proximity between end users and the data center.

11.1.2 User interfaces and access to servers

Cloud computing is extremely technically challenging. However, the difficulties are normally contained by the cloud providers to keep cloud consumers from dealing with technical details. A major strategy is to enable cloud consumers to access cloud services through well-designed user interfaces, which should be intuitive, regardless of the complexity of underlying technology. Cloud consumers should also be able to leverage cloud services through well-defined procedures. This is achieved by using a single-pane management view that consists of tools to help customers go through the life cycle of using cloud computing from configuration and provisioning to management and monitoring (Scheier 2012). This approach potentially makes cloud computing more user-friendly, and enables system administrators to manage a larger pool of resources (Solnik 2012).

11.1.3 Automatic scaling and load balancing

Automatic scaling enables computing resources to be procured or released according to computing demand as predefined by service-level scaling configurations. Dynamic scaling provides a mechanism to maintain system performance when facing fluctuated end user demands (Bondi 2000). A load balancing tool schedules application tasks to available computing resources to meet the performance objectives. Load balancing can optimize the utilization of available resources, provide better computing capacity, and enhance overall system reliability. A load balancer, either software- or hardware-based, is responsible for providing every incoming request with the appropriate computing resource. Different commercial cloud vendors may support different types of scaling and load balancing services. As a result, cloud customers should carefully evaluate solutions from different vendors in order to match their computing requirements with proper cloud services.

11.1.4 Service Level Agreement (SLA)

An SLA formally defines the service contract between the cloud provider and the cloud customer. A typical SLA comprises sections that clearly state different aspects of a service such as what a service is, how the service performance will be evaluated, how to plan crisis management, and how to terminate a service. Generally, an SLA is negotiable between vendors and customers, making it a challenging but crucial part of a cloud computing contract.

Three major cloud services are introduced in the following section. Details of the three services are given for the four considerations when a cloud consumer selects a proper service.

11.2 AMAZON WEB SERVICES (AWS)

11.2.1 Architecture

Amazon is one of the major cloud service providers. One of the services provided by Amazon is Amazon EC2. AWS has a variety of services available to implement different cloud capabilities (Figure 11.1). Within the AWS architecture, there are six layers:

- *Physical infrastructure layer*—Amazon has been steadily expanding its global infrastructure to help its customers achieve lower latency and higher throughput. Currently, Amazon has data centers located in nine regions (Figure 11.2). AWS GovCloud (United States) is an AWS

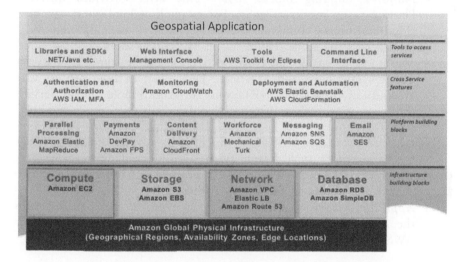

Figure 11.1 AWS General Service architecture. (Revised from Varia, J., 2011, http://www.slideshare.net/AmazonWebServices/amazon-ec2-and-aws-elasticbeanstalk-introduction, accessed on April 28, 2013.)

Figure 11.2 Amazon Global Infrastructure. (From Amazon AWS at http://aws.amazon.com/about-aws/globalinfrastructure/#reglink-na.)

Region designed to allow U.S. government agencies and customers to move more sensitive workloads into the cloud by complying with their specific regulatory and compliance requirements. In addition, Amazon CloudFront and Amazon Route 53 services are offered at AWS Edge locations to deliver content to end users with lower latency (Figure 11.2).

- *Infrastructure building blocks layer*—This layer provides different types of IaaSs to enable cloud customers to access the underlying computing resources (e.g., computing power, storage, and networking). For example, on-demand computing power can be provisioned by Amazon EC2 services; unlimited storage can be offered by Amazon S3 and EBS services, and networking functionalities are provided by Amazon VPC, Elastic LB, and Amazon Route S3.
- *Platform building blocks layer*—This layer provides different Platform as a Services (PaaSs). For example, CloudFront[1] can help cloud consumers configure and deliver content over 21 edge locations across the globe (Figure 11.2); Messaging services allow cloud consumers to easily and cost-effectively program applications with different languages and APIs (e.g., Java, PHP, Python, process vast amounts of Ruby, and .NET); and Amazon Simple E-mail Service (SES) can send bulk and transactional e-mails in a quick and cost-effective manner.
- *Cross service layer*—This layer implements advanced cloud services for management, such as security, monitoring, and automation.
- *Tool and API layer*—Within this layer, different tools and APIs are available to access the underlying AWS.
- *Application layer*—Applications can be migrated and deployed onto the Amazon cloud platform and configured to integrate different AWSs to best serve the cloud consumers.

11.2.2 General characteristics of EC2

11.2.2.1 Scalability

EC2 provides an auto-scaling service.[2] Auto-scaling allows consumers to scale EC2 computing capacity up or down automatically according to predefined conditions, such as response time and CPU utilization. With auto-scaling service, consumers can ensure that the number of EC2 instances used increases seamlessly during demand spikes to maintain performance, and decreases automatically when demand lulls to minimize costs. Auto-scaling is particularly well suited for applications that experience temporal variability in usage.

---

[1] See AWS CloudFront at http://aws.amazon.com/cloudfront/.
[2] See AWS Auto Scaling at http://aws.amazon.com/autoscaling/.

In addition, there are two advanced services in the Cross Service layer of Figure 11.1: Elastic Beanstalk and CloudFront, which can be easily used to implement scalability.

- *Amazon CloudFront* is a Web service for content delivery.[1] It integrates with other AWSs to give cloud consumers an easy way to distribute content to end users with low latency and high data transfer speeds.
- *Elastic Beanstalk*[2] gives cloud customers an even easier way to quickly deploy and manage applications in the AWS cloud without giving up the control of the underlying AWS resources. AWS Elastic Beanstalk automatically handles the details of capacity provisioning, load balancing, auto-scaling, and application health monitoring.

11.2.2.2 Compatibility

Since Amazon EC2 pioneered the cloud service market, its implementations have been generally considered as the de facto standards for different cloud services. Hence, many new cloud providers follow the interface of Amazon EC2.

11.2.2.3 Deployment and interface

AWS provides a variety of tools and APIs to access and manipulate the cloud services. AWS Management Console (Figure 11.3) is a simple and intuitive Web-based user interface for consumers to access and manage AWS. Amazon also provides a companion mobile app for the Android and iOS system to quickly view resources on the fly.

11.2.2.4 Hypervisors

EC2 is a paravirtualized environment based on Xen.[3] It is one of the biggest Xen installations, and uses a heavily modified and adapted version of Xen.

11.2.2.5 Reliability

The Amazon EC2 SLA guarantees 99.95% availability for all Amazon EC2 regions. All other AWS services, such as CloudFront, also provide an SLA with a service level commitment of 99.9% availability in general. For example, Amazon EC2 offers a highly reliable environment for geospatial applications (see Chapters 9 and 10 for details) because the service runs within Amazon's proven network infrastructure and data centers.

---

[1] See AWS CloudFront at http://aws.amazon.com/cloudfront/.
[2] See AWS Elastic Beanstalk beta at http://aws.amazon.com/elasticbeanstalk/.
[3] See On Amazon EC2's Underlying Architecture at http://openfoo.org/blog/amazon_ec2_underlying_architecture.html.

Figure 11.3 AWS Management Console.

11.2.2.6 OS supports

EC2 supports both Windows and Linux operating systems running on its VMs.

11.2.2.7 Cost

EC2 has flexible and multiple price models, which allow cloud consumers to reduce costs based upon workloads. The costs are calculated based on factors such as the tenant model, regions, and computing usage, instance type and operating system of the instances. The tenant model includes (1) On-Demand, (2) Reserved (Light, Medium, and Heavy for 1-Year, 3-Year), and (3) Spot.

- The *On-Demand Instances* model allows consumers to pay for computing capacity by the hour without long-term commitments. Figure 11.4 shows the on-demand instance cost for the U.S East Region (Section 11.2.1).
- The *Reserved Instances* model gives consumers the option to make a low, one-time payment for each instance they reserve and in turn receive a significant discount on the hourly charge for that instance.
- The *Spot Instances* model enables consumers to bid for Amazon EC2 computing capacity. Cloud consumers simply leverage spare Amazon EC2 instances and run them whenever the bid exceeds the current Spot price, which varies in real time based on supply and demand.

Region: US East (N. Virginia) ⌄	Linux/UNIX Usage	Windows Usage
Standard On-Demand Instances		
Small (Default)	$0.065 per Hour	$0.115 per Hour
Medium	$0.130 per Hour	$0.230 per Hour
Large	$0.260 per Hour	$0.460 per Hour
Extra Large	$0.520 per Hour	$0.920 per Hour
Second Generation Standard On-Demand Instances		
Extra Large	$0.580 per Hour	$0.980 per Hour
Double Extra Large	$1.160 per Hour	$1.960 per Hour
Micro On-Demand Instances		
Micro	$0.020 per Hour	$0.020 per Hour
High-Memory On-Demand Instances		
Extra Large	$0.450 per Hour	$0.570 per Hour
Double Extra Large	$0.900 per Hour	$1.140 per Hour
Quadruple Extra Large	$1.800 per Hour	$2.280 per Hour
High-CPU On-Demand Instances		
Medium	$0.165 per Hour	$0.285 per Hour
Extra Large	$0.660 per Hour	$1.140 per Hour
Cluster Compute Instances		
Quadruple Extra Large	$1.300 per Hour	$1.610 per Hour
Eight Extra Large	$2.400 per Hour	$2.970 per Hour
High-Memory Cluster On-Demand Instances		
Eight Extra Large	$3.500 per Hour	$3.831 per Hour
Cluster GPU Instances		
Quadruple Extra Large	$2.100 per Hour	$2.600 per Hour
High-I/O On-Demand Instances		
Quadruple Extra Large	$3.100 per Hour	$3.580 per Hour
High-Storage On-Demand Instances		
Eight Extra Large	$4.600 per Hour	$4.931 per Hour

Figure 11.4 On-demand instance cost for different OS instances in the U.S. east region in February 2013. (Cloud service cost is changing frequently.)

The Spot Instance pricing model complements the On-Demand and Reserved Instance pricing models, providing potentially the most cost-effective option for obtaining computing capacity.

11.2.3 Major users and general comments

11.2.3.1 List of major customers

According to the 451 Group (a leading syndicated research, advisory, and professional services firm), Amazon took approximately 59% of the cloud

Figure 11.5 Amazon EC2 major customers. (See Amazon AWS at http://www.slideshare. net/AmazonWebServices/the-total-cost-of-non-ownership-in-the-cloud.)

service market, while all other cloud service vendors shared the rest of the 41% of cloud consumers in 2011.[1] Amazon EC2 is now powering many popular Internet businesses, trusted by many enterprises and governmental agencies (e.g., NASA and IBM), and has a variety of Association of Southeast Asian Nations (ASEAN) customers (Figure 11.5). Successful applications and businesses supported by EC2 can be found on the Web site.[2]

11.2.3.2 Maturation[3]

Launched in July 2002, AWS provides online services for Web sites or client-side applications. In 2004, Amazon foresaw it would be beneficial to offer infrastructure level service to its customers. As a result, Amazon S3 was launched in March 2006, and Amazon EC2 was built in August 2006 with the Amazon infrastructure and developers base available worldwide. Offering services in 2006 and since then, AWS became the market leader of cloud computing, by virtue of its early entry, rapid innovation, and flexible cloud services. In June 2007, Amazon claimed that more than 330,000 developers had signed up to use AWS. As a core part of AWS, EC2 provides the computing facility for organizations and is capable of supporting a variety of applications (Figure 11.5) (Chapter 5). In November 2010, Amazon made the switch of its flagship retail Web site to EC2 and AWS.

11.2.3.3 Feedback from the community

EC2 offers a variety of cloud services, and has the biggest cloud market share. The biggest concern about EC2 is the potential power outage that could take out the whole regional service center.

---

[1] See Alog at http://www.alog.com.br/en/noticias/infraestrutura-elastica-e-atraente-ao-orcamento/.
[2] See AWS at http://aws.amazon.com/solutions/case-studies/.
[3] See Amazon.com Basic Case Study at http://www.slideshare.net/davinken/amazon-Web-services-history-overview-presentation.

- On June 14, 2012, the Amazon U.S. East data center, its largest data center, lost power because of severe weather conditions. Heroku, Quora, Parse, and Pinterest were among the sites affected by the outage, along with many small companies that rely on Amazon as their source of computing power instead of a traditional data center.[1]
- On June 29, 2012, AWS suffered another outage because of violent electrical storms, taking down many Web site customers.
- On October 22, 2012, Amazon EC2 cloud infrastructure in the eastern United States (North Virginia data center) had a power outage that impacted many Web services including Reddit, Airbnb, Flipboard, GetGlue, Coursera, and more.[2]
- On December 24, 2012, EC2's outage made millions of Netflix customers unable to stream video on a high-demand night for movie viewing.

These failures significantly impact customers' faith on relying exclusively on cloud computing technology, and have caused some customers to quit cloud services (e.g., Whatsyourprice). But Amazon supporters say incidents simply show the need to use multiple availability zones.[3]

11.2.4 Usage complexity

AWS offers user-friendly Web tools for easy access and managing cloud services (Figure 11.1). Figure 11.3 shows the AWS services accessible through the Web console. AWS also provides SDKs, IDE Toolkits, and Command Line Tools for developing and managing consumer applications.[4] With SDKs, AWS services can be integrated in the applications development using different programming languages (e.g., Java, Python, PHP, Ruby, or.Net) or platforms (e.g., Android or IOS). AWS development can also be speeded up with specialized IDE cloud toolkits integrated into the development environment (e.g., Eclipse and Visual Studio). In addition, cloud consumers can control the AWS services from the command line and automate service management with scripts.

---

[1] See Information Week at http://www.informationweek.com/cloud-computing/infrastructure/amazon-defended-after-june-14-cloud-outa/240002207.
[2] See Venture Beat at http://venturebeat.com/2012/10/22/amazon-cloud-outage-takes-down-reddit-airbnb-flipboard-more/.
[3] See Information Week at http://www.informationweek.com/cloud-computing/infrastructure/amazon-defended-after-june-14-cloud-outa/240002207.
[4] See AWS Tools at http://aws.amazon.com/tools/.

11.3 WINDOWS AZURE

11.3.1 Architecture

Windows Azure is one of the most popular cloud services and is provided by Microsoft. Windows Azure provides both PasS and IaaS. Figure 11.6 shows the architecture and supported features of Windows Azure, which includes computing, data services, networking, application services, commerce,

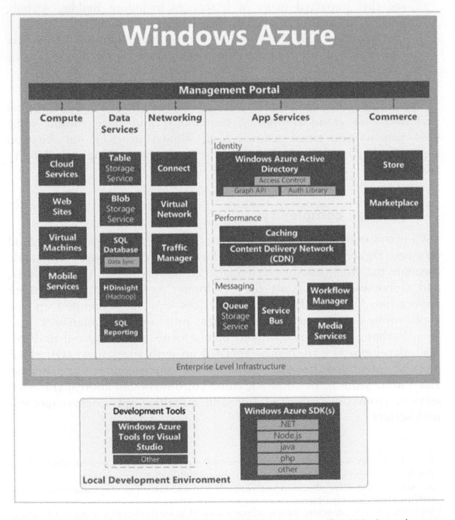

Figure 11.6 The architecture and features of Windows Azure. (See Windows Azure at http://msdn.microsoft.com/en-us/library/windowsazure.)

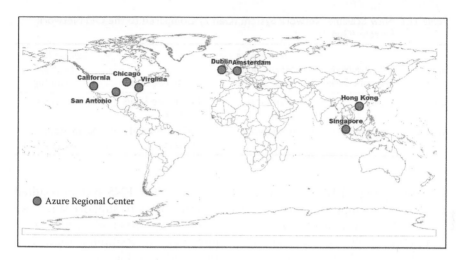

Figure 11.7 Azure Global Infrastructure.

local development environment, and the management portal that access and manage all these components.

Windows Azure provides computing capabilities through four computing components: Web sites, cloud services, virtual machines, and mobile services. At present, there are data centers located in eight regions (Figure 11.7), including U.S. East (Virginia), U.S. West (California), U.S. North Central (Illinois), U.S. South Central (Texas), EU North (Ireland), EU West (Netherlands), Asia East (Hong Kong), and Asia Southeast (Singapore).

Windows Azure has an independent data service, which provides the capabilities to store, manage, and report data. The data services include Windows Azure Table, Blob, and SQL Azure relational database. Windows Azure Networking provides general connections between different components. Window Azure App services provide different modules ensuring identity, performance, and messaging in the cloud platform. Development Tools and Windows Azure SDK provide a local development environment for developers.

11.3.2 General characteristics of Windows Azure

11.3.2.1 Scalability

In Windows Azure, computing resources can be separated from storage to achieve independent scalability of computing and storage. This mechanism also provides isolation and multitenancy. In 2012, the Windows Azure team built a new flat network system that could create Flat Network Storage

Table 11.1 New Scalable Network System (Gen 2) Compared to the Old Network System (Gen 1)

Storage SKU	Storage Node Network Speed	Networking between Computing and Storage	Load Balancer	Storage Device Used for Journaling
Gen 1	1 Gbps	Hierarchical Network	Hardware Load Balancer	Hard Drives
Gen 2	10 Gbps	Flat Network	Software Load Balancer	Solid State Drives (SSDs)

(FNS) crossing all Windows Azure data centers. The FNS system resulted in bandwidth improvements of network connectivity to support Windows Azure VMs. The new networking design better supports high performance computing (HPC) applications that require massive communications and significant bandwidth between computing nodes. Table 11.1[1] shows the improvements of the new network system (Gen 2) from the original network system (Gen 1).

Current Azure storage can provide the following scalability for a single storage account created after June 7, 2012.

- Capacity—Up to 200 terabytes (TBs)
- Transactions—Up to 20,000 entities/messages/blobs per second
- Bandwidth for a georedundant storage account
- Ingress (Chapter 6)—Up to 5 gigabits per second
- Egress (Chapter 6)—Up to 10 gigabits per second
- Bandwidth for a locally redundant storage account
- Ingress—Up to 10 gigabits per second
- Egress—Up to 15 gigabits per second

11.3.2.2 Compatibility

Several Microsoft products have been integrated with Windows Azure to help cloud consumers better use and manage cloud services. For example, consumers can use the Microsoft SQL Server[2] to access and operate (e.g., query, update, delete tables) their SQL Azure database. Microsoft WebMatrix[3] and Visual Studio[4] can be used to remotely access, update, deploy, and configure applications on Windows Azure.

---

[1] See MSDN Blogs at http://blogs.msdn.com/b/windowsazure/archive/2012/11/02/windows-azure-s-flat-network-storage-and-2012-scalability-targets.aspx.
[2] See Microsoft SQL Server at http://www.microsoft.com/en-us/sqlserver/.
[3] See Microsoft WebMatrix at http://www.microsoft.com/web/webmatrix/.
[4] See Visual Studio at http://www.microsoft.com/visualstudio/.

11.3.2.3 Deployment and interface

The Windows Azure interface includes a management portal and power-ful command-line tools. The management portal and command-line tools are easy to use, and help cloud consumers and end users manage applications and cloud services. In addition, several Microsoft products (e.g., Microsoft Visual Studio and Microsoft WebMatrix) can help deploy and manage user applications and cloud resources. Windows Azure also has a REST-based service API for managements and deployments.

11.3.2.4 Hypervisors

Windows Azure Hypervisor is specially designed. This Hypervisor is opti-mized based on a specific homogenous data center environment in Microsoft. Currently, Microsoft does not provide Windows Azure Hypervisor as a single product for cloud consumers and end users.

11.3.2.5 Reliability

Windows Azure guarantees a 99.95% computing reliability and a 99.9% role instance and storage reliability.[1] It provides multiple mechanisms to ensure reliability. For example, it provides a queue mechanism, the Azure Service Bus Queues, to increase the system and application reliability.

11.3.2.6 OS supports

Windows Azure currently supports both Windows and Linux Operation Systems for VMs.

11.3.2.7 Cost

Windows Azure does not have upfront costs. Cost is calculated based on com-puting usage. The cost includes: (1) Web site cost, (2) virtual machines cost, (3) cloud service cost, (4) mobile service cost, and (5) data management cost. The cost for each component is based on the cloud configuration. Windows Azure payment plans include: (1) pay-as-you-go, (2) 6-month plans, and (3) 12-month plans. Microsoft provides an official cost calculator[2] for Windows Azure so that users can easily estimate the cost. Figure 11.8 shows an example of using the Windows Azure price calculator for Web site cost estimation. It must be pointed out that the Windows Azure's price changes dynamically. For example, Windows Azure's cost increased by 5% on February 1, 2013.

---

[1] See IT Strategists at http://www.itstrategists.com/Microsoft-Azure.aspx.
[2] See Windows Azure Pricing Calculator at http://www.windowsazure.com/en-us/pricing/calculator.

Figure 11.8 Windows Azure price calculator.

11.3.3 Major users and general comments

11.3.3.1 List of major customers

Windows Azure has been widely used in various industries and communities. Australian retailer *Harvey Norman* used Windows Azure cloud service to reduce the risk and capital costs for an online bargain site. Hogg Robinson Group extended its core travel platform to mobile devices using Windows Azure Mobile Service. The utilization of Windows Azure has saved up to 80% in startup costs and improved productivity by 25%. The University of Washington moved the Michelangelo (a self-service reporting system) to Windows Azure to address scalability issues of the system. Figure 11.9 shows a part of Windows Azure's major customers.

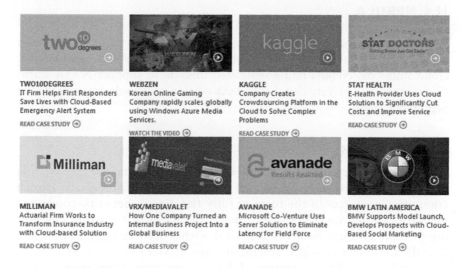

Figure 11.9 Windows Azure major customers. (See Windows Azure Case Studies at http://www.windowsazure.com/en-us/home/case-studies/.)

11.3.3.2 Maturation

Windows Azure was announced on October 2008 followed by SQLAzure Relational Data's announcement on March 2009. Since initiated, Windows Azure has been updated several times to provide more services and functions. In October 2010, full-IIS support and extra small instances have been added to Windows Azure. In December 2011, Traffic Manager, SQL Azure reporting, and HPC scheduler have been supported.

11.3.3.3 Feedback from community

Windows Azure is one of the most popular commercial cloud services. It provides both PaaS and IaaS types of cloud service. Windows Azure attracts many consumers by its compatibility with the .NET environment. Windows Azure's strategic movement to welcome the open-source community potentially adds to its market share in the future.

11.3.3.4 Usage complexity

Windows Azure provides an easy-to-use management portal and powerful command-line tools to manage applications and cloud services. Usage complexity can be even lowered if customers go with the Windows Azure's PaaS service, which saves them from deploying the entire computing environment.

11.4 NEBULA

Nebula is a federal cloud computing service served by NASA. Nebula is listed as an open-source project. However, unlike general open-source cloud solutions, Nebula had a dedicated mission to support NASA scientific tasks with a private cloud. The beta version of Nebula was released in September 2010.

11.4.1 Architecture

Figure 11.10 presents the architectural components of the Nebula platform. IaaS provided by Nebula enables cloud customers to launch computing resources in the form of fully functional, network-connected virtual machines on demand. A unique requirement for cloud computing to support scientific tasks is to deal with a massive amount of scientific data. Nebula provides several storage services that can be used to store and retrieve large amounts of data, at any time, from anywhere on the Web.

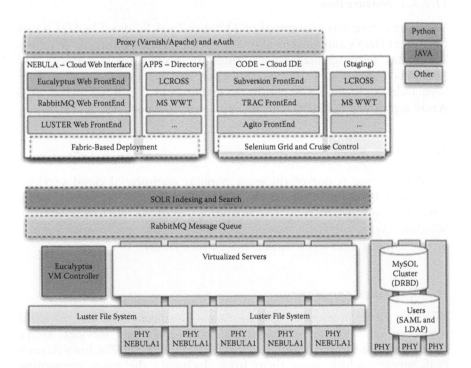

Figure 11.10 Nebula architectural components.

11.4.2 General characteristics of Nebula

11.4.2.1 Scalability

Nebula envisioned supporting cloud consumers for scientific research with a highly scalable, reliable, secure, fast, and inexpensive infrastructure. Eventually, Nebula hopes to achieve the capability of auto-scaling that allows customers to automatically scale cloud services up or down according to defined conditions. Saini et al. (2012) demonstrated that because of application restrictions Nebula has so far realized only limited scalability.

11.4.2.2 Compatibility

From its origin, Nebula was designed to provide services that are compatible with those that are provided by Amazon EC2. Nebula leverages the Eucalyptus API for VM control and acquires compatibility with Amazon EC2 platforms and its clones.

11.4.2.3 Deployment and interface

Nebula provides a graphical user interface and a set of command-line tools for consumers to manage the virtual machines, network access permissions, and so forth.

11.4.2.4 Hypervisors

Nebula uses the Kernel-Based Virtual Machine (KVM) hypervisor.

11.4.2.5 Reliability

As for now, Nebula is still an experimental project. Reliability may be improved as the software is gradually developed.

11.4.2.6 OS supports

Nebula supports two Linux distributions, Ubuntu and CentOS.

11.4.2.7 Cost

Nebula is open-source software, which does not have a cost for the software itself. But cost may be incurred by operations. However, no public information about Nebula's operational cost is available yet.

11.4.3 Major users and general comments

11.4.3.1 List of major customers

Nebula is one of the first cloud solutions built for the federal government by the federal government. The White House was listed as the first customer when Nebula was launched publicly.

11.4.3.2 Maturation

Nebula is still at the early stage, and has been experiencing significant improvements. In several pioneering projects that have implemented Nebula's service, good system performance and satisfying results have been reported.

11.4.3.3 Feedback from community

The U.S. Chief Information Officer praised Nebula for helping NASA to effectively engage the public in scientific activities.

11.4.3.4 Usage complexity

The procedure to set up the initial environment of Nebula and launch the instances is complicated. But the steps to add more nodes for cluster scaling are straightforward. As Nebula is still at its early stage, manual intervention is required from time to time when an instance fails to boot.

11.5 CONCLUSION

As two examples of matured commercial cloud services, both Amazon EC2 and Windows Azure present satisfactory supports for features such as scalability and reliability. NASA Nebula as a private cloud in its internal implementation receives good user experiences and shows promising supports to advanced scientific computations. Amazon cloud services and Windows Azure both have service centers at multiple geographical locations. Both of them have been in the cloud service market for several years, and have been improving their service qualities. NASA Nebula was developed to access the feasibility of adopting a cloud computing model for federal computing. Compared to its commercial peers, Nebula has the flexibility to support unique requirements, such as safety, from the agencies. Table 11.2 outlines comparable aspects of the three platforms. Scalability, reliability, and load balancing have significant impacts on geoscience applications.

Table 11.2 Comparison of the Readiness Considerations for the Three Platforms

	Amazon EC2	Windows Azure	NASA Nebula	Description
Geographical Distribution of Centers	Centers in 9 geographical regions	4 centers in the United States, 2 centers in the EU, and 2 centers in Asia	400 Nebula accounts across 9 NASA centers, the Jet Propulsion Lab, and NASA headquarters	—
Scalability	Auto-scaling service	Computation and storage independent scalability	Limited scalability	Dynamic adjustment of computing resources against tasks
Compatibility	Amazon cloud platform standards are supported by many open-source solutions	No information found	Compatible with Amazon EC2	The readiness of products to be used by different cloud providers
Deployment Interface	AWS Management Console, Mobile app for Android	Management portal, CLI, REST-based API	A GUI and a set of command-line tools.	Installation interface
Hypervisors	Xen	Specially designed for Windows Azure	KVM	Supported virtual machine managers
SLA Defined Reliability	SLA guarantees 99.95% availability for all Amazon EC2 regions	99.95% computing reliability and 99.9% role instance and storage reliability	No information found	Fault tolerance mechanisms in the cloud implementation

11.6 SUMMARY

This chapter introduces three cloud services in detail, and provides supportive knowledge for the next chapter. Section 11.1 introduces core factors that need to be considered when selecting cloud services. Section 11.2 introduces Amazon Web Services (AWS). Section 11.3 introduces Windows

Azure. Section 11.4 introduces NASA Nebula. Section 11.5 summarizes the status of the three cloud services.

11.7 PROBLEMS

1. Please enumerate five key considerations in the commercial cloud service comparison.
2. Please enumerate the six layers of Amazon AWS architecture and briefly summarize their core functionalities.
3. What are the core advantages and the biggest concerns of Amazon AWS?
4. Please enumerate the core components of Windows Azure and briefly summarize their core functionalities.
5. What are the core advantages and the biggest concerns of Windows Azure?
6. What are the major differences of NASA Nebula from the other two cloud services regarding their targeted customers?
7. Enumerate two other popular cloud services and discuss the advantages of using the cloud services?

REFERENCES

AWS. 2012. Overview of Amazon Web Services. http://media.amazonWebservices. com/AWS_Overview.pdf (accessed April 28, 2013).

Bondi, A.B. 2000. Characteristics of scalability and their impact on performance. *ACM WOSP'00 Proceedings of the 2nd International Workshop on Software and Performance*, New York, NY.

ISACA. 2012. Cloud Computing Market Maturity: Study Results. https:// downloads.cloudsecurityalliance.org/initiatives/collaborate/isaca/2012-Cloud-Computing-Market-Maturity-Study-Results.pdf (accessed January 24, 2013).

Li, A., X. Yang, S. Kandula, and M. Zhang. 2010. CloudCmp: Comparing Public Cloud Providers. *IMC'10*, November 1–3, 2010. Melbourne, Australia.

Saini, S., S. Heistand, H. Jin, J. Chang, R. Hood, P. Mehrotra, and R. Biswas. 2012. An application-based performance evaluation of NASA's Nebula cloud computing platform. *2012 IEEE 14th International Conference on High Performance Computing and Communications*, pp. 336–343. June 25–27. Liverpool, UK.

Scheier, R. L. 2012. Virtualization Management: A Single Pane of Glass. *CIO. IN*. http://www.cio.in/article/virtualization-management-single-pane-glass (accessed April 19, 2013).

Solnik, R. 2012. Cloud Management through a Single Pane of Glass, Not a Kaleidoscope. http://www.informationweek.com/cloud-computing/infrastructure/cloud-management-through-a-single-pane-o/240001622 (accessed April 28, 2013).

Varia, J. 2011. Amazon EC2 and AWS Elastic Beanstalk Introduction. http://www.slideshare.net/AmazonWebServices/amazon-ec2-and-aws-elastic-beanstalk-introduction (accessed April 28, 2013).

Yang, C., H. Wu, Q. Huang, Z. Li, and J. Li. 2011. Using spatial principles to optimize distributed computing for enabling the physical science discoveries. *Proceedings of the National Academy of Sciences* 108, 14: 5498–5503.

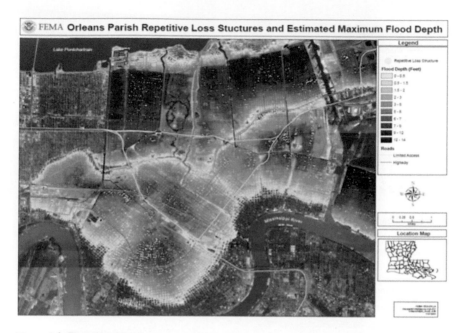

Figure 1.1 Flooding depth map of New Orleans after Hurricane Katrina. (Courtesy of the Federal Emergency Management Agency [FEMA], Department of Homeland Security.)

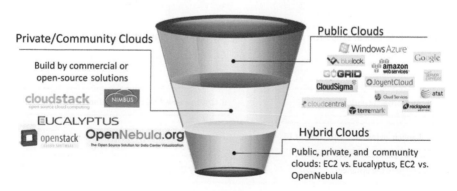

Figure 2.5 Cloud types and software solutions.

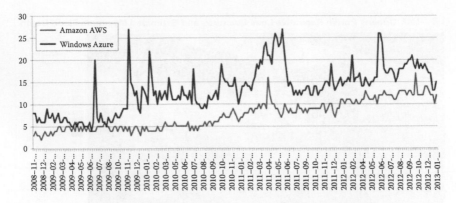

Figure 4.1 Search trends among Amazon Web Services (AWS) and Windows Azure.

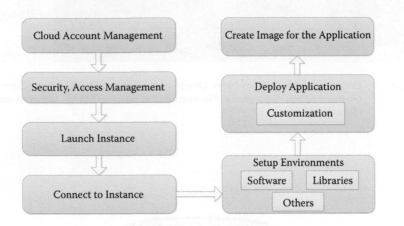

Figure 5.1 General steps for cloud-enabling geoscience applications.

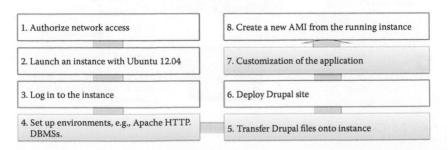

Figure 5.2 The procedure for deploying the Drupal site onto EC2 (blue boxes indicate the additional steps for the deployment).

1. Authorize network access	8. Run the DEM interpolation
2. Launch a cluster of instances as the head node	7. Transfer the DEM data and interpolation code to the head node
3. Install the middleware packages, e.g., Condor	6. Configure the middleware on both nodes to enable communication
4. Create a new AMI from the running instance	5. Start another instance from the new AMI as a computing node

Figure 5.9 The process for configuring an HPC system to run the DEM interpolation on EC2 (blue boxes indicate the additional steps for configuring a virtual HPC environment).

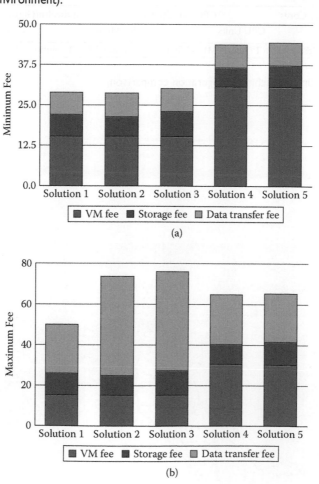

Figure 6.5 (a) Minimum Fee and (b) Maximum Fee charts.

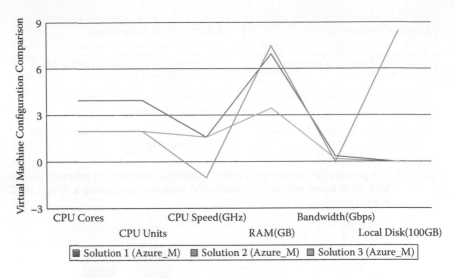

Figure 6.6 Virtual machine configuration comparison.

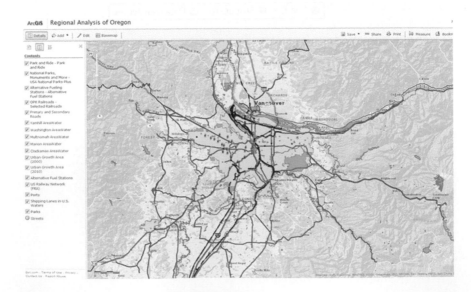

Figure 7.1 Regional Analysis of Oregon map.

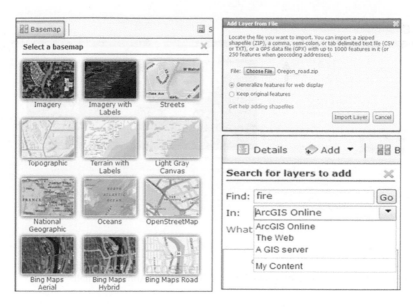

Figure 7.2 Base maps and data offered by ArcGIS Online.

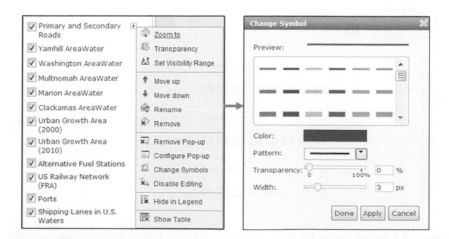

Figure 7.3 Symbol modification of the Primary and Secondary Roads.

Figure 8.6 Search results for global "Annual Sum NDVI Annual Rainfall."

Figure 8.7 Search results for global "Annual Sum NDVI Annual Rainfall" from the GEO Portal.

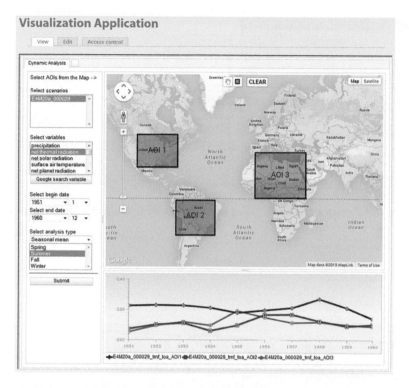

Figure 9.6 Seasonal mean analysis using model outputs.

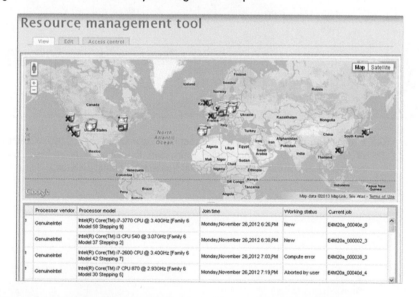

Figure 9.7 Visualizing volunteered computing nodes with Google Maps.

Figure 10.4 Create a placement group and place the instances within the same group.

Figure 10.5 Photograph of a haboob that hit Phoenix, Arizona, on July 1, 2007. (Courtesy of Osha Gray Davidson, *Mother Jones*, San Francisco, CA at http://www. motherjones.com/blue-marble/2009/09/australias-climate-chaos.)

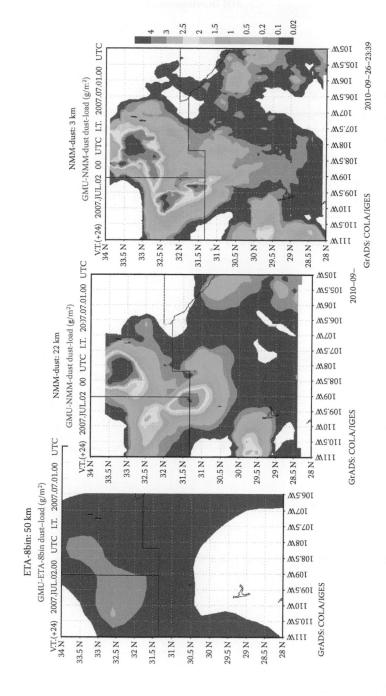

Figure 10.6 Comparison of the simulation results by ETA-8bin and NMM-dust AOI I0, I1, I2, and I3 at 3 A.M., July 2, 2007.

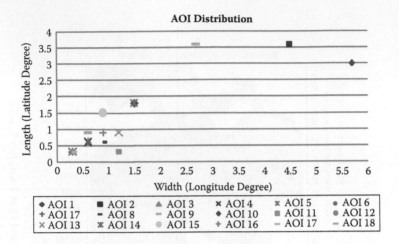

Figure 10.7 AOI distribution identified by the ETA-8bin.

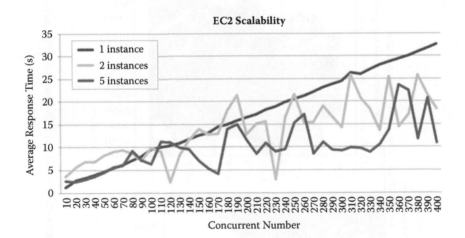

Figure 12.4 EC2 scalability with up to 1, 2, and 5 instances.

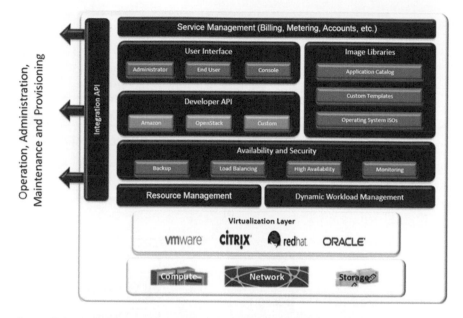

Figure 13.2 Architecture of CloudStack. (Adopted from CloudStack Architecture at http://www.slideshare.net/cloudstack/cloudstack-architecture.)

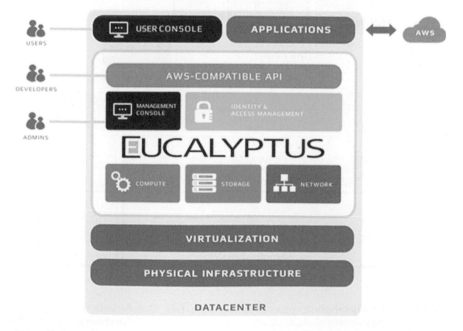

Figure 13.3 Architecture of Eucalyptus. (See Eucalyptus at http://en.wikipedia.org/wiki/File:Eucalyptus_Platform_Architecture,_February_2013.jpg.)

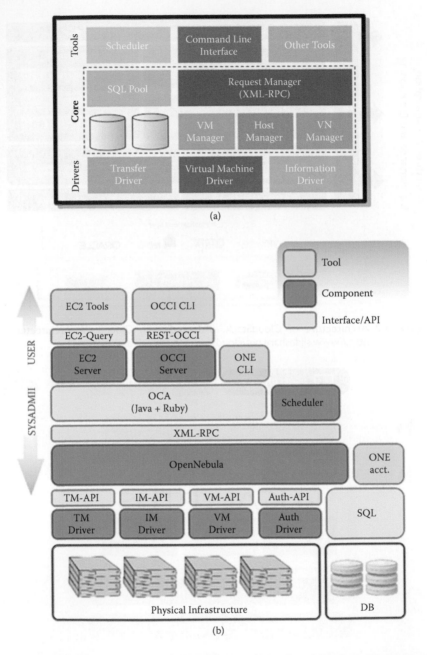

Figure 13.4 (a) OpenNebula internal architecture (see OpenNebula architecture at http://opennebula.org/documentation:archives:rel2.0:architecture) and (b) interfaces (see OpenNebula scalable architecture at http://opennebula.org/documentation:rel3.8:introapis).

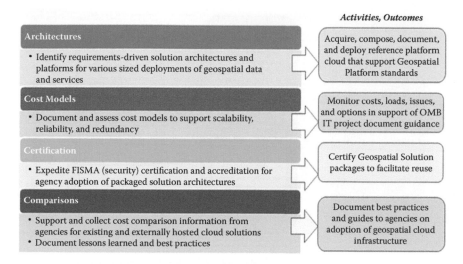

Figure 15.1 GeoCloud Goals, Activities, and Outcomes.

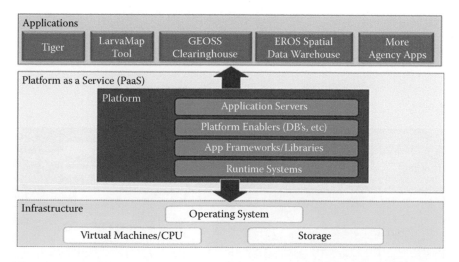

Figure 15.2 GeoCloud architectural framework.

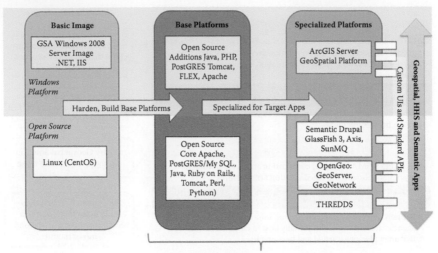

Figure 15.4 GeoCloud platform creation and customization for federal geospatial applications. (From Nebert, 2010, www.fgdc.gov/ngac/meetings/december-2010/geocloud-briefing.pptx.)

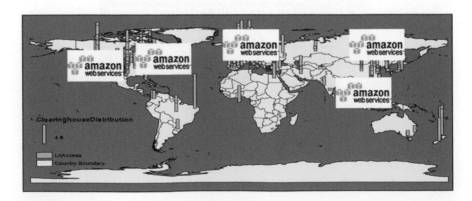

Figure 16.5 Global content delivery to handle global user access.

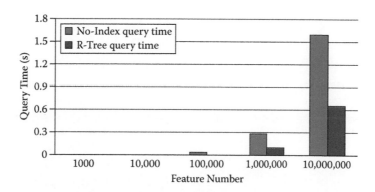

Figure 16.7 Spatial indexing to improve the access speed.

Chapter 12

How to test the readiness of cloud services

Chaowei Yang, Min Sun, Jizhe Xia, Jing Li, Kai Liu, Qunying Huang, and Zhipeng Gui

This chapter utilizes three geoscience applications to demonstrate how to test the readiness of cloud services including NASA Nebula, Windows Azure, and Amazon Elastic Cloud Computing (EC2). The test results from cloud-hosted applications are compared to those using a traditional cluster.

12.1 INTRODUCTION

The flourishing of cloud services provides different options to support geoscience applications. The decision to migrate applications onto cloud services is partially based on the readiness of the services. According to the computational characteristics of geoscience applications (Chapters 1 and 5), the readiness is tested using GEOSS Clearinghouse (CLH), Climate@Home, and Dust Storm Forecasting. This chapter illustrates how to test cloud services with a practical approach by evaluating the performance of exemplar projects on the four different computing services of EC2, Azure, Nebula, and a traditional cluster. A systematic test and analyses of readiness of cloud services to support geosciences is detailed by Yang et al. (2013).

12.2 TEST ENVIRONMENT

The test is conducted on different cloud services located at different geographic regions (Table 12.1) with various network connections and computing configurations.

12.2.1 Network

The three cloud services and the cluster are located at different places (Table 12.1) and connected through the National LamdaRail.[1] End-to-end

---

[1] See National LamdaRail at http://www.nlr.net/.

Table 12.1 Locations of Computing Services

Computing Service	Geographic Location	Cloud Services	Host Organization
EC2	Reston, VA	IaaS	Amazon
Azure	Chicago, IL	PaaS	Microsoft
Nebula	Ames, CA	IaaS	NASA
Local Cluster	Fairfax, VA	Traditional Cluster	George Mason University (GMU)

Table 12.2 Computing Instance Hardware Configuration

Service Names	CPU Cores	CPU Speed (GHz)	Memory (GB)
EC2 1	1	2.3	0.5
EC2 2	2	2.3	7.5
EC2 3	4	2.3	7.5
EC2 4	8	2.8	7.5
Azure 1	1	1.6	1.75
Azure 2	2	1.6	3.5
Azure 3	4	1.6	7
Azure 4	8	1.6	14
Nebula 1	1	2.9	2
Nebula 2	2	2.9	4
Nebula 3	4	2.9	8
Nebula 4	8	2.9	16
Local Server	8	2.4	23

testing is conducted for the performance of each computing service. The concurrent intensity is tested using the clearinghouse by issuing different numbers of concurrent requests to the Clearinghouse running on each computing service. The cloud services are also tested using the computing intensive Climate@Home and dust storm forecasting models.

12.2.2 Computing service configuration

The computing service instances are configured based on the available hardware configuration (Table 12.2).

12.3 CONCURRENT INTENSITY TEST USING GEOSS CLEARINGHOUSE (CLH)

12.3.1 Clearinghouse requirements for computing services

CLH is a FGDC, GEO, and NASA collaborative project to support the sharing of Earth Observation (EO) data in the global context. CLH provides a harvesting function to collect metadata of EO data from remote catalogs, which have been registered in GEOSS, and search capabilities to share the metadata among global users. To support the harvest-share mechanism, CLH uses a variety of geospatial standards including CSW, SRU, and RSS for search and discovery, ISO 19139 for metadata, and WMS, WCS, and WFS related Open Geospatial Consoritum (OGC) standards for data access and visualization. Global user access to the Clearinghouse has different spatiotemporal patterns for different regions and different metadata. For example, the Clearinghouse receives more accesses at local daytime than during evening hours, and the access can be massive when a significant event (e.g., earthquake) happens.

12.3.2 Test design

To compare the capabilities of different cloud services in supporting concurrent requests, two experiments using CLH are used to compare the average response time for concurrent access:

1. *Matrix of concurrent search requests test.* With CLH running on the four computing services, three cloud services (AWS, Azure, and Nebula) and a local server (Figure 12.1), concurrent requests are issued from one service to all the other services. The Apache JMeter[1] is used for testing the performance of concurrent requests between the services. The test result is a 4*4 matrix, which can represent the concurrent performance of each computing service against other computing services.

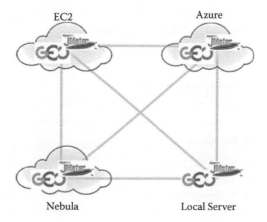

Figure 12.1 Design of a matrix test.

---

[1] See The Apache Software Foundation at http://jmeter.apache.org/.

2. *Load balance and scalability tests.* Load balancing and the auto-scaling mechanism can help maintain good performance by using multiple instances dynamically. The load balancing experiment compares concurrent performance when different numbers of instances are used behind the load balancer; the auto-scaling experiment tracks the change in performance when cloud services scale new instances dynamically with the increasing number of concurrent requests.

12.3.3 Test workflow

1. *Matrix test of concurrent search requests.* Figure 12.2 shows the workflow of the matrix test with the following detailed steps:
   - *Step 1. Installing CLH on different cloud services*—There are two ways to install CLH: (a) Launching an instance using virtual image, the AMI image used for Amazon AWS can be accessed from the AWS Marketplace[1] by searching "CLH." (b) Launching an instance and installing CLH (refer to Chapter 8).
   - *Step 2. Setting up a CSW GetRecords request*—A CSW GetRecords request can be issued to the CLH (refer to Chapter 8). Generally, the request is an XML file and can be kept (e.g., `post.xml`) on the machine, which hosts JMeter. Appendix 12.1 shows a GetRecords example to search for water records.

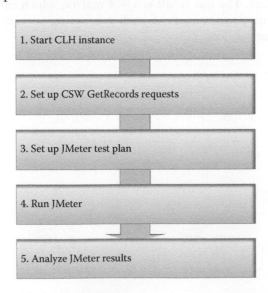

1. Start CLH instance

2. Set up CSW GetRecords requests

3. Set up JMeter test plan

4. Run JMeter

5. Analyze JMeter results

Figure 12.2 The workflow of a matrix test.

---

[1] See AWS Marketplace at https://aws.amazon.com/marketplace.

- *Step 3. Setting up JMeter test plan*—Testers can set a test plan to simulate the concurrent requests to CLH. The concurrent number can be increased, such as 1, 50, 100, 150, 200, 250, 300, 350, 400, 450, 500, and 600. Generally, two types of configurations should be set: thread group and controller. Thread group saves the basic information of the test such as: number of threads, which is used to simulate the number of concurrent requests; ramp-up period, which means the total time to send the full number of requests (five is recommended in the test); and loop count, which means the number of times to execute the test (three is recommended in the test). Besides the thread group, the controller should be set up to keep the information of the CSW requests. Testers should choose HTTP request to set the controller. The following should also be set in the controller: server IP, port, path and request type (POST), the path of the post xml, and the path to save the results file. Appendix 12.2 shows an example of the test plan, which simulates 100 requests to CLH in 1 second.
- *Step 4. Running JMeter*—There are two ways to run JMeter: Graphic User Interface (GUI) and command line. GUI running can be achieved by clicking Run >Start if JMeter is installed with GUI. Meanwhile, we can also run JMeter through the command line. The following script is used to run different test plans that have different numbers of threads. Testers need to go to the bin folder of JMeter and run the following command.

```
# Create the array of the threads number
threads=(1 50 100 150 200 250 300 350 400 450 500 600
700 800 900 1000)
# Run the test using jmeter from this machine to GMU
server
for (( i = 0; i <=11; i++))
do
  for (( j = 0; j <=2; j++))
do
  head=../jmx/GEOSSTestPlan_
  tail=.jmx
  planName=$head${threads[i]}$tail
  ./jmeter -n -t $planName
  sleep 20
done
done
```

- *Step 5. Analyze JMeter results*—This step is introduced in Section 12.3.4.

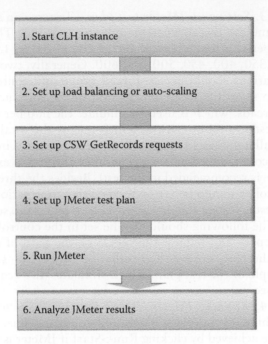

Figure 12.3 The workflow of a load balancing and auto-scaling test.

2. *Load balancing and auto-scaling tests* (Figure 12.3). These two tests can be used to test the improvements of the performance when using load balancing or auto-scaling for CLH. Testers can test stepwise to use 10 to 400 numbers of concurrent requests to do the tests. In addition, testers can also set up how many instances will be brought up in the test, for example, 1, 2, and 5 instances.

Figure 12.3 shows the workflow of the load balancing/auto-scaling test. Steps 1, 3, 4, 5, 6 can be referred to as the *matrix test*. Step 2 sets up the load balancing or auto-scaling as detailed in Chapter 8. CPU utilization is recommended to be used as the resource trigger. For example, testers can set the scalability file to add a new instance once the average CPU utilization in a 300-second period is larger than 90% and to remove an instance once the average CPU utilization in a 300-second period is less than 75%. The detailed configuration can be found in the online material for Chapter 12.

12.3.4 Test result analyses

The JMeter records the response information in the result file as being set up in Step 3. The results file has the basic information of each thread such as response status and response time. Testers can compare the response times

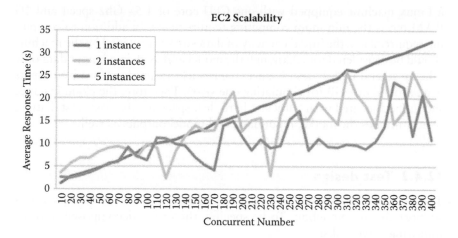

Figure 12.4 (See color insert.) EC2 scalability with up to 1, 2, and 5 instances.

using the results file. For example, Figure 12.4 shows the average response time of GEOSS using Amazon EC2 to scale different levels of concurrent requests. Three groups of experiments are conducted: no scale up and scale down with just one instance, automatically scale up with up to 2 instances, and automatically scale up to 5 instances. The rule to scale up an instance is when the response/latency time is longer than 4 seconds. According to Figure 12.4, using 2 and 5 instances to scale the concurrent access can decrease the average response time. After 50 to 60 concurrent requests, if the average response time is longer than 4 seconds, group "2 instances" and "5 instances" started a new instance. Because it takes time to start the CLH, the average time starts to decrease when the concurrent number is 80. The Amazon AWS log file shows a group of "5 instances" start the third instance 10 minutes later after the second instance starts, the fourth instance is 8 minutes later after the third instance starts, and the fifth instance is 9 minutes later after the fourth instance starts. Hence, there are some drops when the concurrent number is 170, 210, and 270 for the group is "5 instances."

12.4 DATA AND COMPUTING INTENSITIES TEST USING CLIMATE@HOME

12.4.1 Climate@Home computing requirements

The Climate@Home project is to create a virtual supercomputer for large-scale climate modeling and simulations utilizing computing resources from citizen contributors (Chapter 9). ModelE was chosen as the climate model. With the serial processing mode enabled, the minimum hardware requirements of executing the model are quite low (Chapter 9). Generally speaking, a Linux machine equipped with one CPU core of 1.5+ Ghz speed and 1G RAM meets the minimum computing requirements. In addition to computing requirements, the initial capacity of data storage provided by the machine should be considered too. Although the model and associated input data only take up several megabytes of space, the model can produce massive outputs (e.g., around one GB for simulating one year). Therefore, the cloud instance should be allocated with sufficient hard disk space for the basic installation, which includes the model, input data, and model outputs as well.

12.4.2 Test design

The objective of this test is to demonstrate how different cloud services support climate modeling. To be specific, the test is decomposed into the following three tasks:

1. Identifying a cost-effective hardware configuration of the instance through comparing the time costs of executing model runs with virtual instances of different hardware configurations. Because ModelE is a Linux-based application, therefore, only Amazon EC2 or Nebula (supports Linux instances) are used for comparison. A set of sample instances with different numbers of CPU cores and comparable RAM are created to examine the configuration of cloud instances impacting the execution time of model runs (Dai et al. 2009; Onisick 2011).
2. Evaluating the stability and reliability of virtual machine instances by recording the total failure times and average time spent on executing multiple runs. Multiple model runs with different simulation time periods (for example, one day, one month) are executed on every instance. This process is completed while executing a shared script that can control the total number of model runs and change the simulation time periods of each model run.
3. Comparing the performance of cloud-based virtual machines and physical machines. Virtual machine instances have the same hardware configurations as physical machines. The same model runs are executed on both virtual instances and the physical machine.

12.4.3 Test workflow

Based on the design of the tests, the test workflow is defined as follows (Figure 12.5):

- *Step 1. Configuring the computing environment*—Considering the computing requirements of different model runs and available resources, we compiled a list of hardware configurations and created instances

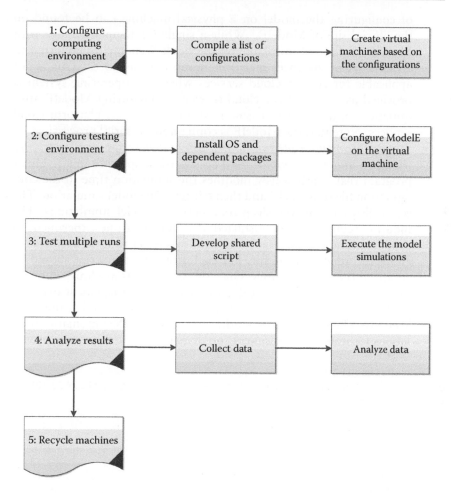

Figure 12.5 The workflow of the experiment for the Climate@Home model simulations.

based on these configurations. In this case, the physical machine is equipped with an 8-core CPU and 8G RAM. Correspondingly, at least one instance should have the same hardware configuration. The CPU configurations of other instances can be 1-core, 2-core, and 4-core.

- *Step 2. Configuring the testing environment*—The same type of operating system (OS) is chosen for all machines, which can minimize the difference among testing results caused by different OSs. The OS should also satisfy the special requirements of compiling ModelE (e.g., gcc version). For every machine, the ModelE should be installed and configured properly. The configuration of ModelE on instances is quite similar to the one on physical machines, except that an instance needs to be accessed through remote control. The detailed procedure of configuring the model on a physical machine can be found on the official site of ModelE.[1] While a single OS template image with ModelE configured (available on the book Web site) is preferred to create a set of instances given the number of instances. This is not applicable for certain cloud services where the operating system is provided as a part of the cloud services. Configuring ModelE after initiating an instance is a generic strategy for cross-platform cloud services. Note that the ModelE is configured for serial processing.

- *Step 3. Testing multiple runs with a Linux shell script on all machines (instance and the computer)*—The shared script contains a Linux program that continuously modifies the simulation time in the configuration file of ModelE and then triggers the model simulation. The script also controls the sleep time between model simulations. The model execution will be restarted after finishing one experiment to avoid runtime errors. The status and execution time of each model run is recorded in a log file. For each instance, the script should be copied to the root directory. Once the simulations are finished on an instance, the log file is uploaded to the central management machine for further analysis. An example of a script can be found at the book Web site. The script can be modified to facilitate the automatic simulations for multiple experiments on demand.

```
#time:19491208
echo "Begin to run the model at " >logModel19491208.
   txt
cd decks
make setup RUN=E4M20one
cd ..
sleep 30 &
wait
date
sed -i 's/YEARE=1961,MONTHE=1,DATEE=1,HOURE=0/YEARE=1
   949,MONTHE=12,DATEE=8,HOURE=0/g' decks/E4M20one/I
   ##replace simulation time here
date >>logModel19491208.txt
echo "----real start----" >>logModel19491208.txt
date >>logModel19491208.txt
time1=$(date +%s)
./exec/runE_new E4M20one -np 1
date >>logModel19491208.txt
echo "----real end-----" >>logModel19491208.txt
time2=$(date +%s)
echo "Model finished in: " $(($time2-$time1))
>>logModel19491208.txt
```

---

[1] See NASA at http://www.giss.nasa.gov/tools/modelE/HOWTO.html#part0_2.

```
Begin to run the model at
Fri Nov 18 15:58:41 PST 2011
----real start----
Fri Nov 18 15:58:41 PST 2011
Fri Nov 18 16:06:06 PST 2011
----real end-----
Model finished in:   445
```

Figure 12.6 An example of the log file ("logModel19491208.txt", 1-core instance from Nebula).

- *Step 4. Collecting and analyzing log files*—As the log file is stored separately on every instance, the log files are collected from each instance. This involves transferring files from the instance to a computer for integration. The methods of analyzing those results are discussed in Section 12.4.4.
- *Step 5. Cleaning and releasing the instance*—Once the tests are finished, instances should be shut down and saved properly to avoid unnecessary resource usage.

12.4.4 Test result analyses

The configurations of virtual machines include 1-core, 2-core, 4-core, and 8-core instances with comparable main memory (Table 12.2). After executing the shared script on every instance, the time cost of each model run is recorded in a log file tagged with the end time of each simulation period. For example, the file logModel19491208.txt indicates that the log file records the time cost in seconds for the simulation period from 1949/12/01 to 1949/12/08. Below is an example of the log file for a 1-core instance of Nebula (Figure 12.6). The log file records the execution start time, the execution end time, and the time cost of the execution. There are two start times: one is recorded after notifying the machine to prepare for the model run and the other is the time when issuing the execution command. When analyzing the log file, we need to clearly label the association of the log files to their machine configurations, which are not captured in the file name.

12.5 CLOUD TEST USING DUST STORM FORECASTING

12.5.1 Dust storm forecasting computing requirements

Dust storm forecasting involves big geoscience data in model input and output, intensive data processing, and computing (Chapter 10). The periodic phenomena of the simulation of dust storms requires multiple computing resources to execute the simulation in parallel to the same set of intensive computations many times. Advanced computing methodologies and techniques are normally used to address the computing challenges (Yang et al. 2011a). A previous study of dust storm simulation (Huang et al. 2013) found that CPU and networking speed have a significant impact on the simulation performance because of massive data exchange and synchronization (as described in Chapter 10). Therefore, the cloud virtual machines for dust storm forecasting should have a fast CPU speed and be connected with a high bandwidth network.

12.5.2 Test design

The dust storm model is configured on a local high performance computing (HPC) cluster and two different cloud services: Amazon EC2 and NASA's Nebula. Table 12.3 shows the configuration of the three computing services.

In order to examine the performance of a local HPC cluster and the two cloud services, we designed the following experiments:

1. *Different numbers of computing nodes from different services for supporting the same dust storm simulations.* This experiment involves using one, two, four, and eight computing nodes of Nebula, EC2, and GMU HPC clusters to test same simulation task.
2. *The same amount of computing resources from different services to support different simulation tasks.* Two computing nodes are used to run seven simulation tasks on each platform to check if the performance of the different platforms is consistent.
3. *The impact of hyperthreading.* Hyperthreading technology makes a single physical processor appear to be multiple logical processors, and significantly improves performance on computing workloads

Table 12.3 The Configurations of the Three HPC Computing Services

Configuration	EC2 4	Nebula 4	GMU HPC
CPU	2.93 GHz	2.8 GHz	2.33 GHz
Numbers of processors	2	2	2
Numbers of cores/node	8	8	8
Supports hyperthreading	Yes (Resulting in 16 CPU cores in total)	No	No
Total memory	23 GB	16 GB	16 GB
Network bandwidth	10 Gbps	10 Gbps	1 Gpbs
Virtualization	Bare-metal virtualization	Paravirtualization	N/A
Resource arranging strategy	Grouping VMs within closer physical machines	N/A	N/A

(Koufaty and Marr 2003). The main advantage of hyperthreading technology is its ability to improve processor resource utilization (Bulpin and Pratt 2004). To test the impact of hyperthreading, the experiment is designed to test the performance of one EC2 instance before and after shutting down the hyperthreading capability. In this experiment, we used one instance of the EC2 with and without hyperthreading capability to run the same model simulation. The EC2 cluster instance type has quad-core processors with hyperthreading. The EC2 documentation states that the cluster compute instance is assigned two quad-core processors (8 cores total); the processors' hyperthreading capabilities resulted in that benchmark reporting 16 total cores (Table 12.2).

12.5.3 Test workflow

The first test in Section 12.5.2 explores different services through using different numbers of computing nodes to run the same model simulation with the following 10 major steps (Figure 12.7):

- *Step 1. Building dust storm model environment images for cloud service (EC2 and Nebula) and setting up a dust storm environment on a local HPC cluster*—Section 10.3 introduces the steps for how to configure the HPC environment and build the images for a dust storm

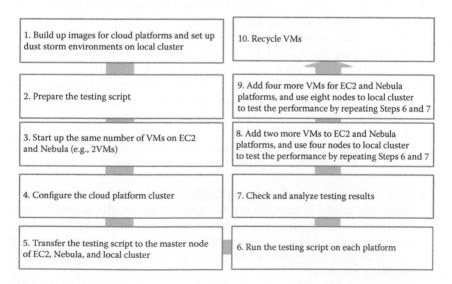

Figure 12.7 The workflow of testing the performance of three different platforms with 2, 4, and 8 VMs using the dust storm model.

model. Building a local HPC cluster environment should follow the same procedures except for building an image, which is not needed for a local HPC cluster.

- *Step 2. Writing the testing shell script for running the model and recording the performance.*

```
#!/bin/sh
## usage: ./run_test.sh [test platform] [VM numbers]
platform=$1
vms=$2
## Run model and record time
cd /model_ directory/model
startTime='date +%s%N'
./run_model.sh
endTime='date +%s%N'
echo 'expr \( $endTime - $startTime \) / 1000000'
## compress and copy model output data back to
repository
tar zcvf $platform.test.$ vms.tar.gz modeloutput
scp $platform.test.$ vms.tar.gz root@199.26.xxx.xxx:/
    state/partition1/
```

- *Step 3. Starting the same number of instances on EC2 and Nebula platforms*—Based on the images created from Step 1, the consumer can launch different numbers of instances. In our test, we will first test the performance of two instances.
- *Step 4. Configuring the cloud platform.* This step is to ensure that the MPICH2 works properly on the VMs.
- *Step 5. Transferring the testing script to the head node of EC2, Nebula, and the local HPC cluster.* The testing scripts written in Step 2 should be copied to the head node of each platform.
- *Step 6. Running the testing script on each platform*—If the script file name is run _ test.sh, then we can use the following command to run the script for the EC2 platform with two instances.

```
[root@headnode~] chmod a+x run_test.sh
[root@headnode~] ./run_test.sh EC2 2VMs
```

- *Step 7. Checking results of each platform and analyzing the results*— Since the model is running on multiple resources, the network communication among those resources may crash sometimes. Therefore, consumers should log in to the head node and check the progress and model output every one or two hours to make sure that the model is running properly.
- *Step 8. Adding two more instances for the EC2 and Nebula platform and use four nodes to test the performance by repeating*

Steps 6 and 7—After the experimental group with two instances is finished, we can start two more instances and form a four-node HPC cluster for both EC2 and Nebula, and add two servers to the local HPC cluster as well. In order to make sure the CPU and memory on the first two instances are released from previous tasks, we can stop the MPICH2[1] services and wait for several minutes. Then we can start to run the next round of simulation with four instances.

- *Step 9. Adding four more instances for EC2 and Nebula platform and use eight nodes to test the performance by repeating Steps 6 and 7*—Repeating Step 8, we can construct eight instances and run the test again.
- *Step 10. Recycling instances*—After the tests are finished, instances can be shut down and released for other applications.

The second experimental test (Section 12.5.2) is to use the same amount of computing resources from different platforms to support different simulation tasks. The test procedure is similar to Figure 12.7, where we should include Steps 1 to 7. In this situation, consumers can start any number of instances (e.g., one, two, or four) to test different problem-size simulations.

In the third hyperthreading technology test (Section 12.5.2), we can start a VM with and without hyperthreading to run the same model simulation task. In default, all EC2 instances have hyperthreading enabled. The following command can be used to disable the hyperthreading for an EC2 instance.

```
for cpunum in $(
    cat/sys/devices/system/cpu/cpu*/topology/thread_
        siblings_list |
cut -s -d, -f2- | tr ',' '\n' | sort -un); do
    echo 0 > /sys/devices/system/cpu/cpu$cpunum/online
done
```

12.5.4 Test result analyses

Figure 12.8 shows a typical test result with different process numbers. The first line ****** Test type EC2.2VMs********** indicates that it is the test result by EC2 cloud service with two instances. The third line indicates the model running directory. The fourth line means the starting time to run the model with 128 process numbers. The fifth line tells the model completing time with 128 processes. The following lines indicate the

---

[1] See MPICH2 at http://phase.hpcc.jp/mirrors/mpi/mpich2/.

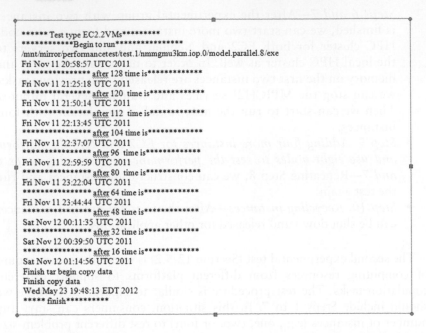

```
****** Test type EC2.2VMs*********
***********Begin to run***************
/mnt/mirror/performancetest/test.1/nmmgmu3km.iop.0/model.parallel.8/exe
Fri Nov 11 20:58:57 UTC 2011
*************** after 128 time is************
Fri Nov 11 21:25:18 UTC 2011
*************** after 120 time is************
Fri Nov 11 21:50:14 UTC 2011
*************** after 112 time is************
Fri Nov 11 22:13:45 UTC 2011
*************** after 104 time is************
Fri Nov 11 22:37:07 UTC 2011
*************** after 96 time is************
Fri Nov 11 22:59:59 UTC 2011
*************** after 80 time is************
Fri Nov 11 23:22:04 UTC 2011
*************** after 64 time is************
Fri Nov 11 23:44:44 UTC 2011
*************** after 48 time is************
Sat Nov 12 00:11:35 UTC 2011
*************** after 32 time is************
Sat Nov 12 00:39:50 UTC 2011
*************** after 16 time is************
Sat Nov 12 01:14:56 UTC 2011
Finish tar begin copy data
Finish copy data
Wed May 23 19:48:13 EDT 2012
***** finish**********
```

Figure 12.8 Model test output on EC2 with two instances and different process numbers.

starting and ending time with other process numbers. The last line with Finish shows the test has been finished successfully.

Through processing the test file output in Figure 12.8, the execution time of a three-hour and 2.3 * 3.5 degree model forecasting by the EC2 cloud service using two instances and different process numbers can be calculated and analyzed (Figure 12.9a). By combining different test output files, more complex analysis can be performed (Figure 12.9b).

12.6 SUMMARY

Geoscience applications can be data, computing, concurrent, and spatiotemporal intensive (Yang et al. 2011b). This chapter describes how to use the three geoscience applications, CLH, Climate@Home, and dust storm forecasting, to test different cloud services. Sections 12.1 and 12.2 introduce the background and test environment. Section 12.3 introduces concurrent intensity testing. Section 12.4 introduces the data and computing intensity test. Section 12.5 introduces comprehensive testing using a dust storm project. The testing method, experiences, and scripts provided through the book's online Web site can be used to test other cloud services. More results and systematic tests are reported by Yang et al. (2013).

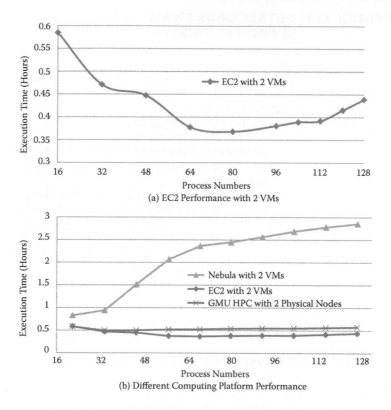

Figure 12.9 Dust storm model performances with different platforms and process numbers.

12.7 PROBLEMS

1. What are the computing characteristics of geoscience applications?
2. Use an example to illustrate how to test the capability of a cloud application in response to concurrent intensity.
3. Use an example to illustrate how to test the capability of a cloud application in response to data intensity.
4. Use an example to illustrate how to test the capability of a cloud application in response to computing intensity.
5. What capabilities of a cloud-hosted application can be tested using the dust storm simulation application?
6. What are the general tools used in the chapter to test the readiness of cloud services?
7. Read the results paper (Yang et al. 2013) and describe the results in 500 words. Discuss the dynamics of the results, that is, how the results may change.

APPENDIX 12.1: GETRECORDS EXAMPLE TO
SEARCH METADATA

```xml
<?xml version="1.0"?>
<csw:GetRecords xmlns:csw="http://www.opengis.net/cat/
    csw/2.0.2" xmlns:gmd="http://www.isotc211.org/2005/gmd"
    service="CSW" version="2.0.2">
        <csw:Query typeNames="csw:Record">
            <csw:Constraint version="1.1.0">
                <Filter xmlns="http://www.opengis.net/
                ogc"  xmlns:gml="http://www.opengis.
                net/gml">
                    <PropertyIsLike wildCard="%"
                    singleChar="_" escapeChar="\">
                        <PropertyName>AnyText</
                        PropertyName>
                        <Literal>%water%</Literal>
                    </PropertyIsLike>
                </Filter>
            </csw:Constraint>
        </csw:Query>
</csw:GetRecords>
```

APPENDIX 12.2: EXAMPLE OF JMETER TEST PLAN

```xml
<?xml version="1.0" encoding="UTF-8"?>
<jmeterTestPlan version="1.2" properties="2.1">
    <hashTree>
    <TestPlan guiclass="TestPlanGui" testclass="TestPlan"
    testname="GEOSSTestPlan_1" enabled="true">
        <stringProp name="TestPlan.comments">This the for
        the GEOSS Cloud Test</stringProp>
        <boolProp name="TestPlan.functional_mode">false</
        boolProp>
        <boolProp name="TestPlan.serialize_
        threadgroups">true</boolProp>
        <elementProp name="TestPlan.user_defined_
        variables" elementType="Arguments"
        guiclass="ArgumentsPanel" testclass="Arguments"
        testname="User Defined Variables"
        enabled="true">
            <collectionProp name="Arguments.arguments"/>
        </elementProp>
<stringProp name="TestPlan.user_define_classpath"></
stringProp>
</TestPlan>
<hashTree>
```

```xml
<ThreadGroup guiclass="ThreadGroupGui"
testclass="ThreadGroup" testname="Thread_Group_1"
enabled="true">
<stringProp name="ThreadGroup.on_sample_
error">continue</stringProp>
<elementProp name="ThreadGroup.main_controller"
elementType="LoopController"
guiclass="LoopControlPanel"
testclass="LoopController" testname="Loop Controller"
enabled="true">
<boolProp name="LoopController.continue_
forever">false</boolProp>
<stringProp name="LoopController.loops">1</stringProp>
</elementProp>
<stringProp name="ThreadGroup.num_threads">100</
stringProp>
<stringProp name="ThreadGroup.ramp_time">1</
stringProp>
<longProp name="ThreadGroup.start_
time">1319814086000</longProp>
<longProp name="ThreadGroup.end_time">1319814086000</
longProp>
<boolProp name="ThreadGroup.scheduler">false</
boolProp>
<stringProp name="ThreadGroup.duration"></stringProp>
<stringProp name="ThreadGroup.delay"></stringProp>
</ThreadGroup>
<hashTree>
<HTTPSamplerProxy guiclass="HttpTestSampleGui"
testclass="HTTPSamplerProxy" testname="HTTP_Request_
GetRecords" enabled="true">
<elementProp name="HTTPsampler.Arguments"
elementType="Arguments" guiclass="HTTPArgumentsPanel"
testclass="Arguments" testname="User Defined
Variables" enabled="true">
<collectionProp name="Arguments.arguments"/>
</elementProp>
<stringProp name="HTTPSampler.domain">clearinghouse.
cisc.gmu.edu</stringProp>
<stringProp name="HTTPSampler.port">80</stringProp>
<stringProp name="HTTPSampler.connect_timeout"></
stringProp>
<stringProp name="HTTPSampler.response_timeout"></
stringProp>
<stringProp name="HTTPSampler.protocol"></
stringProp>
<stringProp name="HTTPSampler.contentEncoding"></
stringProp>
```

```
<stringProp name="HTTPSampler.path">geonetwork/srv/
en/csw</stringProp>
<stringProp name="HTTPSampler.method">POST</
stringProp>
<boolProp name="HTTPSampler.follow_redirects">true</
boolProp>
<boolProp name="HTTPSampler.auto_redirects">false</
boolProp>
<boolProp name="HTTPSampler.use_keepalive">true</
boolProp>
<boolProp name="HTTPSampler.DO_MULTIPART_
POST">false</boolProp>
<elementProp name="HTTPsampler.Files"
elementType="HTTPFileArgs">
<collectionProp name="HTTPFileArgs.files">
<elementProp name="/root/jakarta-jmeter-2.5.1/bin/
post.xml" elementType="HTTPFileArg">
<stringProp name="File.path">/root/jakarta-
jmeter-2.5.1/bin/post.xml</stringProp>
<stringProp name="File.paramname"></stringProp>
<stringProp name="File.mimetype">application/xml</
stringProp>
</elementProp>
</collectionProp>
</elementProp>
<boolProp name="HTTPSampler.monitor">false</boolProp>
<stringProp name="HTTPSampler.embedded_url_re"></
stringProp>
</HTTPSamplerProxy>
<hashTree/>
<ResultCollector guiclass="TableVisualizer"
testclass="ResultCollector" testname="View Results in
Table" enabled="true">
<boolProp name="ResultCollector.error_
logging">false</boolProp>
<objProp>
<name>saveConfig</name>
<value class="SampleSaveConfiguration">
<time>true</time>
<latency>true</latency>
<timestamp>true</timestamp>
<success>true</success>
<label>true</label>
<code>true</code>
<message>true</message>
<threadName>true</threadName>
<dataType>true</dataType>
<encoding>false</encoding>
<assertions>true</assertions>
```

```
      <subresults>true</subresults>
      <responseData>false</responseData>
      <samplerData>false</samplerData>
      <xml>true</xml>
      <fieldNames>false</fieldNames>
      <responseHeaders>false</responseHeaders>
      <requestHeaders>false</requestHeaders>
      <responseDataOnError>false</responseDataOnError>
      <saveAssertionResultsFailureMessage>false</
      saveAssertionResultsFailureMessage>
      <assertionsResultsToSave>0</assertionsResultsToSave>
     <bytes>true</bytes>
     </value>
     </objProp>
     <stringProp name="filename">/home/result/result.xml</
     stringProp>
     </ResultCollector>
    <hashTree/>
    </hashTree>
  </hashTree>
  </hashTree>
</jmeterTestPlan>
```

REFERENCES

Bulpin, J. R. and I. A. Pratt. 2004. Multiprogramming performance of the Pentium 4 with hyper-threading. In *Workshop on Duplicating, Deconstructing, and Debunking (WDDD04)*, June.

Dai, Y., B. Yang, J. Dongarra, and G. Zhang. 2009. Cloud Service Reliability: Modeling and Analysis. From *The 15th IEEE Pacific Rim International Symposium on Dependable Computing. Hyperthreading*, 2011. http://en.wikipedia.org/wiki/Hyper-threading (accessed April 25, 2013).

Huang Q., C. Yang, K. Benedict, S. Chen, A. Rezgui, and J. Xie. 2013. Utilize cloud computing to support dust storm forecasting. 6(4): 338–355.

Koufaty, D. and D.T. Marr. 2003. Hyperthreading technology in the netburst micro-architecture. *Micro, IEEE* 23, no. 2: 56–65.

Mell, P. and T. Grance. 2009. The NIST Definition of Cloud Computing Ver. 15. http://csrc.nist.gov/groups/SNS/cloud-computing/ (accessed April 25, 2013).

Onisick, J. 2011. Cloud availability 101. *Network Computing*, August.

Yang, C., M. Goodchild, Q. Huang et al. 2011a. Spatial cloud computing—How can geospatial sciences use and help to shape cloud computing? *International Journal of Digital Earth* 4: 305–329.

Yang, C., H. Wu, Q. Huang, Z. Li, and J. Li. 2011b. Using Spatial Principles to Optimize Distributed Computing for Enabling the Physical Science Discoveries. doi: 10.1073/pnas.0909315108.

Yang, C., Q. Huang, J. Xia et al. 2013. Cloud service readiness for geospatial sciences. *International Journal of Geographic Information Science* (in press).

Chapter 13

Open-source cloud computing solutions

*Chen Xu, Zhipeng Gui, Jing Li, Kai Liu,
Qunying Huang, and Myra Bambacus*

In addition to commercial cloud services, many open-source cloud computing solutions can be flexibly tailored to build private cloud services for the specific demands of a user. This chapter introduces four major open-source cloud computing solutions including CloudStack, Eucalyptus, Nimbus, and OpenNebula.

13.1 INTRODUCTION TO OPEN-SOURCE CLOUD COMPUTING SOLUTIONS

While more commercial IT enterprises are providing cloud services through their products (Armbrust et al. 2010; Huang et al. 2012), emerging open-source cloud solutions are capable of transforming the existing infrastructure of an organization to a private or a hybrid cloud service (Sotomayor et al. 2009).

A key challenge IT enterprises face when building a cloud service is managing physical and virtual resources (such as servers, storage, and networks) in a holistic fashion (Rimal et al. 2011). The software toolkit responsible for this orchestration is called a *virtual infrastructure manager* (VIM) (Figure 13.1). Generally, VIMs should provide a set of features including: (1) managing and monitoring the life cycle of virtual machines (VMs), such as creating and releasing a VM, connecting VMs together, and setting up virtual disks for VMs, (2) placing and replacing VMs dynamically on a pool of physical infrastructure, (3) scheduling VMs on physical machines, and (4) providing networking capabilities, such as setting up public and private IP addresses and domain names to enable the VMs to be accessible through a network.

CloudStack, Eucalyptus, Nimbus, and OpenNebula are four of the most popular VIMs available for building many private- and community-based cloud services. The four VIMs are introduced in the following section.

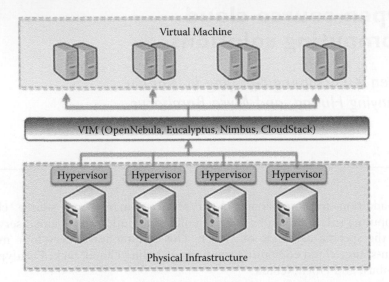

Figure 13.1 The architecture of an Infrastructure as a Service (IaaS) managed by VIM.

13.1.1 CloudStack

CloudStack is an open-source VIM. Multiple hypervisors can be used by CloudStack, such as the three hypervisors introduced in Chapter 3 (VMware ESXi, Xen, and KVM). In addition to its own API, it also implements the Amazon EC2 and S3 APIs to support interoperability with Amazon cloud services. CloudStack was initially developed by cloud.com, a startup supported by venture capital. In late 2011, cloud.com was purchased by Citrix, and CloudStack was donated to the Apache Software Foundation in 2012. Currently, CloudStack is licensed under Apache License, Version 2.

13.1.2 Eucalyptus

Eucalyptus is the abbreviation for Elastic Utility Computing Architecture Linking Your Programs to Useful Systems (Eucalyptus). As an open-source VIM software, Eucalyptus is able to support enterprise private and hybrid cloud computing. Eucalyptus supports hybrid cloud services by integrating Amazon Web Services (AWS) API. Therefore, consumers can create a hybrid cloud service using Eucalyptus for a private cloud and Amazon for a public cloud, and move the VMs between the two clouds. Eucalyptus was developed by researchers at the University of California, Santa Barbara. The first version of the software was released in 2008. The latest version is Eucalyptus 3.2.1[1] released in February 2013.

---

[1] See Download Eucalyptus at http://www.eucalyptus.com/download/eucalyptus.

13.1.3 Nimbus

Nimbus focuses on the computing demands of scientific users. Nimbus has two variations, Nimbus Infrastructure and Nimbus Platform. Nimbus Infrastructure provides a solution for IaaS, and has been designed to support the needs of data-intensive scientific research projects. Nimbus Platform is a set of tools that assist consumers to leverage IaaS cloud computing. The toolset comprises functions for application installation, configuration, monitoring, and repair. Nimbus Platform enables consumers to create hybrid clouds of Nimbus Infrastructure, Amazon AWS, and other clouds. Nimbus supports Xen and KVM hypervisors. Nimbus was developed by the Argonne National Laboratory at the University of Chicago.

13.1.4 OpenNebula

OpenNebula is capable of managing private, public, or hybrid IaaS clouds. OpenNebula users have the flexibility to decide which cloud interfaces are to be adapted from EC2 Query, OGF's Open Cloud Computing Interface (OCCI), or vCloud. Xen, KVM, and VMware hypervisors are supported by OpenNebula. OpenNebula was started as a research project in 2005. The first version was released in 2008. By 2013, there have been more than 5,000 downloads per month.[1] The latest version of OpenNebula is 3.8.3.[2]

The following sections (Sections 13.2 through 13.5) detail the architectures and characteristics of the four open-source cloud solutions.

13.2 CLOUDSTACK

CloudStack is designed to manage large networks of VMs for enabling a highly available and scalable IaaS platform. It includes the entire "stack" of features, which most organizations demand within an IaaS cloud:[3] compute orchestration, Network as a Service, user and account management, a full and open native API, resource accounting, and user-friendly User Interfaces (UI).

13.2.1 Architecture

The CloudStack architecture[4] (Figure 13.2) envisions an IaaS platform with computing, network, and storage resources to be managed as outlined: all resources are tied together through a shared architecture comprising at least one hypervisor solution. CloudStack provides a core segmentation model

---

[1] See OpenNebula at http://c12g.com/open-source-release-of-the-opennebulaapps-suite/.
[2] See OpenNebula Download at http://opennebula.org/software:software.
[3] See Apache CloudStack at http://incubator.apache.org/cloudstack/.
[4] See CloudStack Architecture at http://www.slideshare.net/cloudstack/cloudstack-architecture.

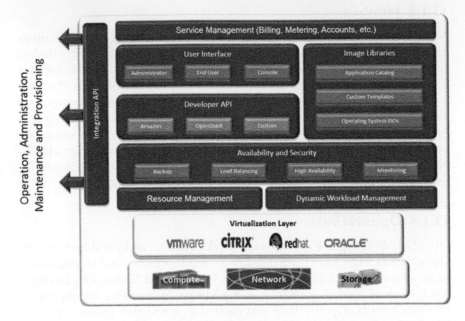

Figure 13.2 (See color insert.) Architecture of CloudStack. (Adopted from CloudStack Architecture at http://www.slideshare.net/cloudstack/cloudstack-architecture.)

based on account management and resource allocation. This model provides a multi-tenant mode for supporting tenancy abstraction ranging from departmental to public cloud reseller modes. It encompasses core functions such as the user interface and image management, and allows cloud providers to provide advanced services such as high availability and load balancing. All services are tied together through a series of Web service APIs, which fully automate CloudStack to support the unique needs of consumers.

13.2.2 General characteristics of CloudStack

- *Scalability*—CloudStack is capable of managing large numbers of servers across geographically distributed data centers through a linearly scalable, centralized management server. This capability eliminates the need for intermediate cluster-level management servers.[1] It supports integration with both software and hardware firewalls and load balancers to provide additional security and scalability to a user's cloud environment, such as F5 load balancer[2] and Netscaler.[3]

---

[1] See Understanding Apache CloudStack at http://incubator.apache.org/cloudstack/software.html.

[2] See F5 Load Balancer at http://www.f5.com/glossary/load-balancer/.

[3] See Citrix at http://www.citrix.com/products/netscaler-application-delivery-controller/overview.html.

- *Cloud model*—CloudStack is a solution for IaaS-level cloud platforms. It pools, manages, and configures network, storage, and computing nodes to support public, private, and hybrid IaaS clouds.
- *Compatibility*—In addition to its own APIs, CloudStack is compatible with Amazon EC2 and S3 APIs,[1] as well as the vCloud APIs for hybrid cloud applications.[2]
- *Deployment and interface*—Users can manage their clouds with a user-friendly Web-based interface, command-line tools, or through a RESTful API. CloudStack provides a feature-rich out-of-the-box user interface that implements the CloudStack API to manage the infrastructure. It is an AJAX-based solution compatible with most popularly used Web browsers. A snapshot view of the aggregated storage, IP pools, CPU, memory, and other resources in use gives users comprehensive status information about the cloud.
- *Hypervisors*—CloudStack is compatible with a variety of hypervisors including VMware vSphere, KVM, Citrix XenServer, and Xen Cloud Platform (XCP).
- *Reliability*—CloudStack is a highly available and scalable IaaS solution, in that no single component failure can cause cluster or cloud-wide outage (Apache 2012). It enables downtime-free management for server maintenance and reduces the workload of managing a large-scale cloud deployment.[3]
- *OS support*—CloudStack supports Linux for the management server and for the underlying computing nodes. Depending on the employed hypervisor, CloudStack supports a wide range of guest operating systems including Windows, Linux, and various versions of Berkeley Software Distributions (BSDs).
- *Cost*—CloudStack itself is free software licensed under the Apache License. However, costs may be incurred by using a commercial hypervisor.

13.2.3 Major users and general comments on the CloudStack solution

- *List of customers*—CloudStack is used by companies to provide a wide range of cloud services from public cloud services and private cloud services, to hybrid cloud services. Many leading telecommunication companies choose Citrix Cloud Platform, powered by Apache CloudStack, as the foundation for their next generation cloud services.

---

[1] See Apache CloudStack Features at http://incubator.apache.org/cloudstack/software/features.html.
[2] See Apache CloudStack at http://en.wikipedia.org/wiki/CloudStack.
[3] See Understanding Apache CloudStack at http://incubator.apache.org/cloudstack/software.html.

These companies include: British Telecom (BT),[1] China Telecom,[2] EVRY,[3] IDC Frontier,[4] KDDI,[5] Korea Telecom (KT),[6] Nippon Telegraph and Telephone (NTT),[7] and Slovak Telecom.[8] Other companies include Tata communications,[9] Nokia Research Center,[10] and Logicworks.[11]

- *Maturity*—CloudStack was first released in 2010, and since then it has been significantly improved. It was first released under the GNU General Public License, Version 3 (GPLv3). When CloudStack was donated to Apache Software Foundation (ASF) by Citrix in April 2012, the license was changed to the Apache License, Version 2. Now, CloudStack is an Apache Incubator Project, Version 4.1.1.[12]
- *Feedback from community*—CloudStack is one of the leaders for open-source cloud solutions, and provides ample functionalities and a user-friendly Web-based GUI. Becoming a subproject of Apache Software Foundation dramatically increased the development speed of CloudStack, and solidified its leadership in open-source cloud solutions.
- *Installation complexities*—CloudStack installation is straightforward. Apache provides a wiki-based site with comprehensive guidelines, various forums, and plenty of documents for different types of users (e.g., general users, administrators, and developers).

13.3 EUCALYPTUS

Eucalyptus is an open-source solution for IaaS cloud computing. Eucalyptus-based IaaS cloud services give consumers the capability to run and control virtual machine instances deployed across a variety of physical resources (Nurmi et al. 2009).

13.3.1 Architecture

Figure 13.3 shows the high-level Eucalyptus architecture. Eucalyptus provides an IaaS solution to build private or hybrid clouds. By virtualization

---

[1] See bt at http://www.bt.com/.
[2] See China Telecom at http://en.chinatelecom.com.cn/.
[3] See Evry at http://www.evry.com/.
[4] See IDC Frontier at http://www.idcf.jp/english.
[5] See KDDI at http://www.kddi.com/english/index.html.
[6] See KT at http://www.kt.com/eng/main.jsp.
[7] See NTT Communications at http://www.ntt.com.
[8] See Telekom at http://www.telekom.sk/.
[9] See Tata at http://instacompute.com/.
[10] See Nokia Research at http://research.nokia.com/.
[11] See Logicworks at http://www.logicworks.net/.
[12] See What is CloudStack at http://incubator.apache.org/cloudstack/.

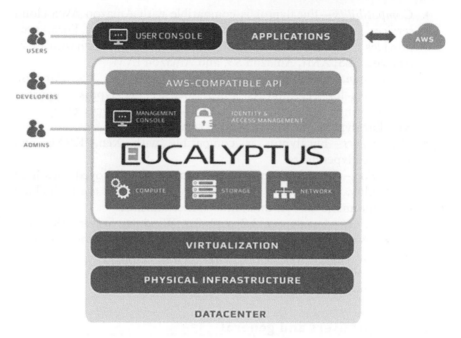

Figure 13.3 (See color insert.) Architecture of Eucalyptus. (See Eucalyptus at http://
en.wikipedia.org/wiki/File:Eucalyptus_Platform_Architecture,_February_2013.jpg.)

of physical machines in the data center, cloud providers can provide collections of virtualized computer hardware resources, including computing, network, and storage to cloud consumers. Consumers can access
the cloud through command-line tools (euca2ools)[1] or through a Web-
based dashboard such as Hybridfox.[2] Eucalyptus also supports an AWS-
compatible API on top of Eucalyptus for consumers to communicate
with AWS.

13.3.2 General characteristics of Eucalyptus

- *Scalability*—Eucalyptus supports scalability starting with
  Eucalyptus 2.0 at two levels: front-end transactional scalability and
  back-end resource scalability (Eucalyptus 2.0 2013).
- *Cloud model*—Eucalyptus provides access to collections of virtualized computer hardware resources for compute, network, and storage
  to provide IaaS cloud services. Users can assemble their own virtual
  cluster where they can install, maintain, and execute their own application stack using Eucalyptus.

---

[1] See Euca2ools at http://www.eucalyptus.com/download/euca2ools.
[2] See Hybridfox at http://code.google.com/p/hybridfox/.

- *Compatibility*—Eucalyptus is compatible with Amazon AWS cloud services, so users can reuse existing AWS tools and scripts. Eucalyptus provides compatibility with a range of AWS features: Amazon EC2, Amazon EBS, AMI, Amazon S3, and Amazon IAM.
- *Deployment and interface*—Eucalyptus supports the Amazon AWS APIs for EC2 and S3. Euca2ools are Eucalyptus supported command-line tools for interacting with Web services compatible with Eucalyptus and Amazon cloud services.
- *Hypervisors*—Eucalyptus is compatible with Xen, KVM, and VMware Hypervisors.
- *Reliability*—Since Version 3, Eucalyptus can be deployed with high availability. Eucalyptus 3 improved the reliability of the IaaS cloud using automatic failover and failback mechanisms.
- *OS support*—Eucalyptus 3.2 supports Windows Server 2003 and 2008, Windows 7, and all modern Linux distributions such as RedHat, CentOS, Ubuntu, Fedora, and Debian.
- *Cost*—Users can choose between the open-source free Eucalyptus Cloud and the priced Eucalyptus Enterprise Cloud.

13.3.3 Major users and general comments on Eucalyptus

- *List of customers*—IT organizations across the globe run Eucalyptus clouds for their agility, elasticity, and scalability required by highly demanding applications (Our Customers 2013). These organizations include: Puma,[1] NASA, National Resource Conservation Service (NRCS), George Mason University, Cornell University, and others.
- *Maturity*—Since Eucalyptus released its first version in 2008, it has released six versions. The latest version, 3.2, was released on December 19, 2012.
- *Feedback from community*—Eucalyptus supports the community with a wiki-based site and mailing lists. Cloud Computing World Series placed it in the Top 3 of "Best Cloud Service category" in 2012.[2]
- *Installation complexities*—Eucalyptus is easy to install and deploy. It provides comprehensive documentation,[3] a responsive community,[4] and mailing lists[5] for consumers and developers.

---

[1] See Puma at http://www.puma.com/.
[2] See Cloud World Series at http://www.cloudcomputinglive.com/awards/2012shortlist.html.
[3] See Eucalyptus at http://www.eucalyptus.com/docs.
[4] See Eucalyptus Engage at https://engage.eucalyptus.com/.
[5] See GitHub Eucalyptus Mailing Lists at http://lists.eucalyptus.com/cgi-bin/mailman/listinfo.

13.4 OPENNEBULA

OpenNebula is designed to provide a solution for building enterprise level data centers and IaaS clouds. Its modular-based architecture allows cloud builders to configure and implement a diverse range of cloud services while maintaining a high level of stability and quality.

13.4.1 Architecture

As one of the most popular open-source cloud computing solutions, OpenNebula is fully capable of enabling private, public, and hybrid cloud platforms (Milojičić et al. 2011). As of Version 2.0, the internal components of OpenNebula include a core module, a set of plug-in drivers, and multiple tools (Figure 13.4a). The core module manages and monitors virtual resources such as VMs, virtual networks, virtual storage, and images. It also handles client requests and invokes corresponding drivers to perform operations on resources. Drivers serve as adapters to interact with middleware. Core functions are exposed to end users through a set of tools and APIs. While the earlier versions of OpenNebula did suffer deficiencies of scalability, which made it only suited for small- and medium-sized clouds (Sempolinski and Thain 2010), its latest version, Version 3.8, has evolved to a highly scalable architecture (Figure 13.4b). The interfaces have been greatly enriched, and encompass REST API (e.g., EC2-Query API), the OpenNebula Cloud API (OCA), and APIs for native drivers. In particular, it has native support for connecting to AWS.

13.4.2 General characteristics of OpenNebula

According to the documentation on OpenNebula,[1] the major features of OpenNebula can be summarized as follows:

- *Scalability*—OpenNebula has been employed in building large-scale infrastructure as well as highly scalable databases. The virtualization drivers can be adjusted to achieve the maximum scalability.
- *Cloud model*—Designed to support IaaS clouds by leveraging existing infrastructure, OpenNebula usually does not have specific requirements on the infrastructure.
- *Compatibility*—OpenNebula can be deployed to existing infrastructure and integrated with various cloud services. To upgrade the system, administrators can follow the compatibility guide in the documentation.[2]

---

[1] See OpenNebula 3.8 Guides at http://opennebula.org/documentation:rel3.8.
[2] See Apache CloudStack at http://incubator.apache.org/cloudstack/docs/en-US/index.html.

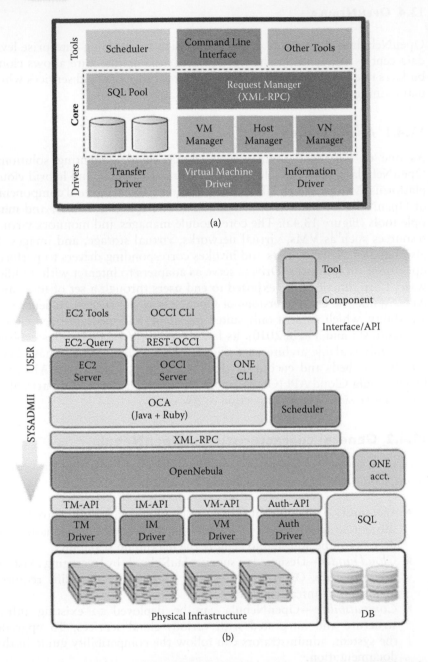

Figure 13.4 *(See color insert.)* (a) OpenNebula internal architecture (see OpenNebula architecture at http://opennebula.org/documentation:archives:rel2.0:architecture) and (b) interfaces (see OpenNebula scalable architecture at http://opennebula.org/documentation:rel3.8:introapis).

- *Deployment and interfaces*—Interfaces are available for cloud consumers and providers. Cloud providers can develop customized tools with cloud interfaces. Consumers can use either the Command Line Interface (CLI) or the SunStone Web Portal to perform most operations, especially the management of resources. In addition, the latest release provides interfaces to cloud providers such as Amazon.
- *Hypervisors*—OpenNebula supports three major hypervisors: KVM, Xen, and VMware. Because the hypervisor driver can be shifted between different hypervisors, it provides a solution for a multihypervisor platform.
- *Reliability*—The OpenNebula system has designed a specialized quality check module, OpenNebula QA, to ensure the quality of every release. Even if the management solution fails accidentally, the virtual resources located on individual machines can be restored.
- *OS support*—Major Linux and Windows versions are supported. The OS images created by hypervisors with or without OpenNebula can be used as the OS templates.
- *Cost*—While OpenNebula is a completely free solution, its enterprise version, OpenNebulaPro, is distributed on an annual subscription basis. Commercial support is a value-added service provided by C12G Labs[1] at a customized cost.

13.4.3 Major users and general comments on OpenNebula

- *Customers*—OpenNebula has been widely employed in various application scenarios as an enterprise-ready solution. Four major categories of users can be identified: (1) Enterprise private clouds and data center virtualization (notable customers include IBM Global Business Services and the National Central Library of Florence); (2) Hosting and cloud service providers such as China Mobile; (3) The High Performance Computing (HPC) science community such as NASA; (4) Cloud integrators and product developers such as the consulting firm KPMG.
- *Maturity*—Since its establishment in 2008, the OpenNebula Project has released more than 20 versions including stable and beta releases as of 2012. The latest version, 3.8, was released on October 22, 2012. Meanwhile, the developers have worked closely with the user community of OpenNebula to address users' requirements as well as technical problems.
- *Feedback from the community*—Feedback from the community also reflects the core values of OpenNebula: "Openness," "Excellence,"

---

[1] See C12G Labs at http://c12g.com/.

"Cooperation," and "Innovation." In particular, cloud providers and consumers emphasize the openness that allows a high degree of customization to applications. More feedback can be found at OpenNebula Testimonials at http://opennebula.org/users:testimonials.

- *Installation complexities*—OpenNebula relies on other open-source software such as KVM and it does not offer an official integrated installer. Before installing the software, users have to install all dependent packages and configure the environment accordingly. Therefore, the installation is relatively easy for experienced installers who are familiar with Linux and virtualization techniques. In contrast to experienced users, first-time users may experience a steep learning curve.

13.5 NIMBUS

The initial development of the Nimbus system targeted computing requirements of scientific studies. Nimbus comprises a set of tools for establishing IaaS cloud service.

13.5.1 Architecture

Nimbus comprises the Nimbus Platform and the Nimbus Infrastructure. The Nimbus Platform lets users use IaaS clouds, and Nimbus Infrastructure supports users to build IaaS clouds. Figure 13.5 illustrates the components of the Nimbus Infrastructure. When a cloud consumer is subscribed to a

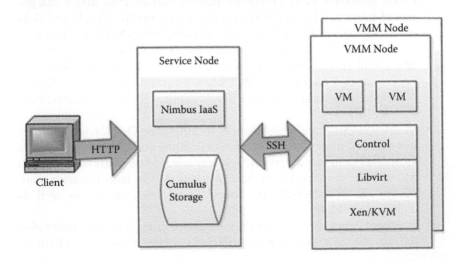

Figure 13.5 Nimbus IaaS structure.

Nimbus service, a virtual workspace is created. The workspace comprises the front end, the workspace service, the back end, and the VM workspace. The VM workspace is deployed onto the Virtual Machine Monitor (VMM) node, which is a physical node. Once the deployment has been done, consumers can access the cloud service node via the HTTP interface. Cumulus is a crucial component of Nimbus, serving as the front end to the Nimbus VM image repository. Any VM image must be loaded into the Cumulus repository before booting.

13.5.2 General characteristics of Nimbus

- *Scalability*—The Cumulus Redirection module of Nimbus manages scalability. This module keeps track of the workload of the service. As Cumulus is compatible with the Amazon S3 REST API, it can be configured to run as a set of replicated hosts to support horizontal scalability.
- *Cloud model*—Nimbus is a VIM solution for IaaS.
- *Compatibility*—Cumulus storage extends the Amazon S3 REST API, and is S3 compatible.
- *Deployment and interface*—Users directly interact with VMs in the node pool in almost the same way as interacting with a physical machine. Nimbus publishes information about the VM such as the IP address of each VM so that users can know information about each VM. Users deploy applications to Nimbus clouds by using a *cloudkit* configuration that includes a manager service hosting and an image repository.
- *Hypervisors*—Nimbus supports KVM and Xen.
- *Reliability*—To achieve the same level of reliability as S3, the hardware configuration of Cumulus needs to be at the same level as S3. In other words, the reliability of Nimbus partially depends on the hardware infrastructure the Cumulus builds on.
- *OS support*—Nimbus supports various Linux distributions.
- *Cost*—Nimbus is an open-source solution. Therefore, there is no cost for the software.

13.5.3 Major users and general comments on Nimbus

- *Customers*—Nimbus was first developed for scientific tasks. It has been used for the STAR project at the Brookhaven National Laboratory, ING, JPMorgan, and Toyota.
- *Maturity*—Nimbus started as a research project in 2003. The latest stable version is 2.10.[1]

---

[1] See Nimbus at http://www.nimbusproject.org/downloads/.

- *Feedback from the community*—Nimbus can meet the needs of scientific computing communities (Mangtani and Bhingarkar 2012).
- *Installation complexity*—The installation of Nimbus begins with installing a set of software to one service node in a nonroot account. The Nimbus cloud client must be installed on additional VMMs.

13.6 OPEN-SOURCE BENCHMARKING CONSIDERATIONS

Tables 13.1 and 13.2 summarize the comparison among the four IaaS solutions based on considerations that should be given when comparing different options. Table 13.1 profiles architectural characteristics, and Table 13.2 describes user setups.

13.7 CONCLUSION

The open-source community has provided mature solutions for establishing IaaS cloud services. CloudStack, Eucalyptus, Nimbus, and OpenNebula are four solutions that have been successfully implemented by IT enterprises as well as by academic research institutes. Compared to cloud services by commercial providers, open-source solutions offer additional flexibility and better control of resources. For customers who have considerations of custodianship and security, open-source cloud computing solutions provide options for establishing private clouds. The four open-source IaaS solutions introduced above each have the capacity to integrate different types of hardware and software. Due to differences in managing the underlying hardware (e.g. CPU, RAM, etc.), the operational performance of each solution might be different. Therefore, platform performance needs to be tested for supporting various computing tasks (from data intensive, computing intensive, communication intensive to concurrent intensive tasks) and will be covered in Chapter 14.

13.8 SUMMARY

This chapter introduces four IaaS open-source cloud computing solutions in general (Section 13.1) as well as detailed facts for: CloudStack (Section 13.2), Eucalyptus (Section 13.3), OpenNebula (Section 13.4), and Nimbus (Section 13.5). Based on the details of each solution, a comparison of the four solutions has been drawn on their system characteristics (Section 13.6). Contents in this chapter lay the foundation for the next chapter, which benchmarks the four solutions.

Table 13.1 Comparisons of the Four IaaS Solutions

	Nimbus	Eucalyptus	CloudStack	OpenNebula
Scalability (Capability for managing)	Scalable	Scalable	Scalable	Scalable
Compatibility (Compatibility with other cloud services, e.g., AWS)	EC2, S3 Compatible	Support EC2, S3	Amazon EC2 (through CloudBridge)	Open, multiplatform
Deployment (The capability of adding and deleting physical nodes)	Dynamic deployment	Dynamic deployment	Dynamic deployment	Dynamic deployment
Installation Interface (Interface for installing the package)	CLI	CLI	CLI	CLI
Disk Image Options (Create an image from scratch or rebundle)	Depends on configuration	Options set by service providers	Users are able to upload and manage templates and images.	In private cloud, most libvirt options left open
Hypervisors (Supported virtualization software)	KVM, Xen	VMWare (in nonopen source, Xen, KVM	VMware, KVM, Oracle VM, Kronos, Citrix XenServer, and Citrix Xen Cloud Platform	Xen, VMWare, KVM
Web Interface (Connection protocols)	RESTful HTTP API	Web service	Ajax- and jQuery-based Web GUI,	
Secure VNC console access, support API from Amazon and VMware	XML-RPC API, EC2 Query, OGF OCCI, and vCloud APIs			
Structure (The way system components such as client, agent, and resource manager are organized)	Lightweight components	Modular	Modular	Modular

(Continued)

Table 13.1 (Continued) Comparisons of the Four IaaS Solutions

	Nimbus	Eucalyptus	CloudStack	OpenNebula
Reliability (Fault tolerance mechanisms in the cloud implementation)	Depends on the underlying hardware	Yes, using automatic failover and failback mechanisms	Yes, multiple management server and database server; and also VMs live-migration functionalities	Rollback host and VM
Built-in Monitoring (VM, cloud infrastructure monitoring)	Cloudinit.d	VM state	VM sync and high availability	Basic variables
Firewall Management (Managing accessibility and firewall of VMs)	ebtables filtering tool for a bridging firewall	Security groups	User group to manage instance firewall	Use software router for firewall management
Load balancer (Providing load balancing services)	No, can be augmented by load balancers such as Zeus	Zeus Load Balancer	Using other traffic balancer service for load balancer	Yes, uses built-in load balancer plugin

Table 13.2 Defined User Actions and Interfaces

	Nimbus	Eucalyptus	CloudStack	OpenNebula
User Permissions	Admin, user	Admin, user	Admin, user	Admin, user by group
Instance Connection	SSH	SSH keypairs, RDP for Windows instances	VNC, SSH	VNC, SSH
Cloud Resource Control	CLI and Web-based CMS	CLI, Euca2ools, and EC2 API	Support CLI and API from Amazon and VMware	CLI, Java API, Web-based CMS

13.9 PROBLEMS

1. What are the general features a virtual infrastructure manager (VIM) should provide?
2. What are the general characteristics of CloudStack?
3. What are the general characteristics of Eucalyptus?
4. What are the general characteristics of OpenNebula?
5. Please describe the major differences between Nimbus and other VIMs introduced in the chapter.
6. Use an example to illustrate which one of the three VIMs, CloudStack, Eucalyptus, and OpenNebula, will be the most suitable solution for the example.

REFERENCES

Apache. 2012. CloudStack Installation Guide. http://incubator.apache.org/cloudstack/docs/en-US/Apache_CloudStack/4.0.0-incubating/pdf/Installation_Guide/Apache_CloudStack-4.0.0-incubating-Installation_Guide-en-US.pdf (accessed March 25, 2013).

Armbrust, M., A. Fox, R. Griffith et al. 2010. A view of cloud computing. *Communications of the ACM 53*, no. 4: 50–58.

Eucalyptus. 2013. Eucalyptus 2.0 Delivers High Scalability and Flexibility for Private Cloud Computing. http://www.eucalyptus.com/news/08-25-2010 (accessed August 8, 2013).

Huang, Q., J. Xia, C. Yang et al. 2012. An experimental study of open-source cloud platforms for dust storm forecasting. In *Proceedings of the 20th International Conference on Advances in Geographic Information Systems*, pp. 534–537. ACM.

Mangtani, N. and S. Bhingarkar. 2012. The appraisal and judgment of Nimbus, Open Nebula and Eucalyptus. *International Journal of Computational Biology 3*, no. 1: 44–47.

Milojičić, D., I. M. Llorente, and R. S. Montero. 2011. OpenNebula: A cloud management tool. *Internet Computing IEEE* 15, no. 2: 11–14.

Nurmi, D., R. Wolski, C. Grzegorczyk et al. 2009. The Eucalyptus open-source cloud-computing system. In *Cluster Computing and the Grid. CCGRID'09. 9th IEEE/ACM International Symposium*, pp. 124–131.

Rimal, B. P., A. Jukan, D. Katsaros, and Y. Goeleven. 2011. Architectural requirements for cloud computing systems: An enterprise cloud approach. *Journal of Grid Computing* 9, no. 1: 3–26.

Sempolinski, P. and D. Thain. 2010. A comparison and critique of Eucalyptus, OpenNebula and Nimbus. In *Cloud Computing Technology and Science (CloudCom), IEEE 2nd International Conference*, pp. 417–426.

Sotomayor, B., R. S. Montero, I. M. Llorente, and I. Foster. 2009. Virtual infrastructure management in private and hybrid clouds. *Internet Computing, IEEE* 13, no. 5: 14–22.

Chapter 14

How to test the readiness of open-source cloud computing solutions

Qunying Huang, Jizhe Xia, Min Sun, Kai Liu, Jing Li, Zhipeng Gui, Chen Xu, and Chaowei Yang

This chapter introduces how to test and compare the readiness of three open-source cloud computing solutions, including Eucalyptus, CloudStack, and OpenNebula.

14.1 INTRODUCTION

The objective of this chapter is to test the readiness of open-source solutions in supporting the NIST cloud computing characteristics and computing requirements for geosciences. The test matrix includes performance, usability, costs, and characterizations of the solutions. Initially, three cloud solutions are selected for the test: Eucalyptus, Cloudstack, and OpenNebula. Factors determining the selection of the open-source solutions include: (i) maturity of the solutions, (ii) availability of an open-source version, (iii) cost of the enterprise edition, (iv) size of the user community, and (v) any publicly available quality indication (market share, feedback of customers, and others). Tables 14.1 and 14.2 summarize our findings including the key customers of the cloud solutions, feedback from the community on the maturity of the solutions, installation complexities, availabilities of an open-source version, and cost. Supported Operating Systems (OS) for physical machines and virtual machines (VMs) are also listed.

The focus of this chapter is on operational knowledge and a detailed presentation and discussion of the testing results are provided by Huang et al. (2013).

14.2 TEST ENVIRONMENT

The cluster built and hosted by the Center of Intelligent Spatial Computing for Water/Energy Science (CISC) (Figure 14.1) is dedicated to providing the testing environment. Each open-source cloud computing solution is

Table 14.1 Customers and Evaluations of the Software Solutions

Solution	Partial List of Customers	Maturation (Years in Business)	Feedback from Community	Installation Complexities
Eucalyptus	Cloudera, DoD, NASA, HP, SONY, FDA, Puma, USDA	The first version was released in May 2008	While the software is easy to install, cloud consumers have reported challenges of network configuration	Installation is relatively simple for first-time installer; Network configuration is difficult
OpenNebula	IBM, Hexagrid, CloudWeavers, CERN	The first public release was in 2008	New stable version has been contributed to the project	Steep learning curve for first-time installer
CloudStack	Zynga, Edmunds.com, Nokia Research Center	CloudStack was released in May 2010	Well documented; May have problems with creating instances with Xen	Easy to install because packages are prepared for guiding installers through the process

Table 14.2 The General Characteristics of the Selected Solutions

Solution	Open-Source Version	Cost (Enterprise Edition)	OS That Can Support the Cloud	OS Images That Can Be Supported
Eucalyptus	Yes	Licensing based on number of physical hosts	Red Hat Enterprise; CentOS; openSUSE-11; Debian; Fedora; Ubuntu	Windows; Major Linux OS distributions
OpenNebula	Yes	Free	Ubuntu and CentOS	Windows; Major Linux OS distributions
CloudStack	Yes	Free	Red Hat Enterprise; CentOS 5; CentOS 6.0; CentOS 6.1; Ubuntu 10.04; Fedora 14	Windows; Major Linux OS distributions

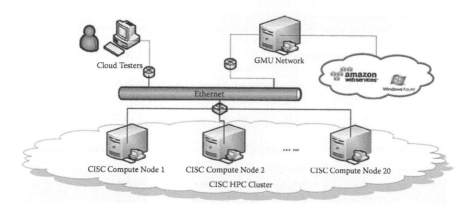

Figure 14.1 The open-source solution test environment.

installed and configured on the same hardware infrastructure environment. Twenty-six computing nodes are utilized from the CISC cluster and connected through local area networks (LANs) (with 1 Gbps). Each node has a 16 GB memory size and a dual quad-core 2.33 GHz central processing unit (CPU).

Figure 14.2 illustrates the conceptual architecture of the open-source solutions (Eucalyptus, CloudStack, and OpenNebula). The bottom layer has the same hardware configuration and network connection setup for all three solutions. The operating system (OS) layer includes CentOS 5.7, CentOS 5.8, and CentOS 6.0 as the OS. Xen (Barham et al. 2003) or Kernel-Based Virtual Machine (KVM) (2010), which serve as the virtualization layer managed by the open-source software. Each solution is installed on five computing nodes with one serving as the master node and the other four serving as the computing (or slave) nodes. The application layer includes two geoscience applications for evaluating application performance.

Five VM configurations are tested on each of the three solutions (Table 14.3). Each VM configuration has different random access memory (RAM) size and virtual CPU unit cores.

14.3 TESTS OF CLOUD OPERATIONS

Performances of single computing resource allocation and release are tested. The start and release of computing resources, including VM and storage volume, are tested. Each test is repeated five times for each VM configuration. Time costs for starting, pausing, restarting, and

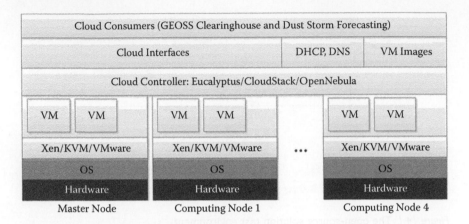

Figure 14.2 Conceptual cloud architecture.

Table 14.3 Cloud VM Configurations

Instance Type	Architecture	OS	RAM (GB)	Virtual Cores	Disk (GB)
Small	64-bit	CentOS 6.0	1.7	1	10
Medium	64-bit	CentOS 6.0	1.7	2	10
Large	64-bit	CentOS 6.0	7.5	2	10
XLarge	64-bit	CentOS 6.0	12	4	10
XXLarge	64-bit	CentOS 6.0	12	8	10

deleting a VM with different virtual computing resources are recorded and compared.

14.4 TESTS OF VIRTUAL COMPUTING RESOURCES

14.4.1 Brief introduction

The virtual computing resource tests include VM CPU, memory, hard drive, and networking performance benchmarking. Table 14.4 outlines the selected benchmark toolsets for the tests. UBench tests CPU performance. Bonnie++ is an open-source benchmarking tool for evaluating hard drive and file system performance. CacheBench, a component of the LLCbench package (Mucci 2012), evaluates the performance of a computer system's memory hierarchy. Iperf (Tirumala et al. 2012) measures Transmission Control Protocol (TCP) and User Datagram Protocol (UDP) bandwidth performance.

The following sections introduce the virtual computing resource test design, workflow, and output interpretation.

Table 14.4 General Benchmarkers for Testing the Performance of Virtual Machines (VMs)

	Benchmarker	Test items
Virtual Computing Resource	UBench[a]	CPU
	Bonnie++ [b]	Hard drive
	CacheBench[c]	Memory Hierarchy Performance
	Iperf[d]	Network
General Application	sql-bench	MySQL
	SPECjvm2008[e]	Java
Geoscience Application	Spatial Web portal (database, tomcat)	GEOSS Clearinghouse
	MPI application	Dust storm model

[a] From UBench, 2012, http://www.tucows.com/preview/69604/Ubench.
[b] From Coker, 2008, http://www.coker.com.au/bonnie++/.
[c] See CacheBench at http://icl.cs.utk.edu/projects/llcbench/cachebench.html.
[d] From Tirumala et al., 2012, http://iperf.sourceforge.net/.
[e] See Spec at http://www.spec.org/download.html.

14.4.2 Test design

- *CPU performance.* The C++ based UBench (UBench 2012) benchmarking tool is used to measure the CPU performance. Ubench executes a series of floating-point and integer calculations for 3 minutes per test.
- *I/O performance.* Bonnie++ (Coker 2008) is used for benchmarking I/O performance by testing file system operations to provide a result of the amount of work completed per second and the percentage of CPU time the operations took. Two indicators are used for I/O benchmarking: (1) the performance results with higher numbers are better and (2) CPU usage with lower percentage is better. File operations that Bonnie++ performs are: (1) data read and write speed, (2) number of seeks per second, and (3) number of file metadata operations per second.
- *Memory hierarchy performance.* LLCbench (Low Level Characterization Benchmark Suite) (Mucci 2012) was created by combining MPBench,[1] CacheBench1,[2] and BLASBench[3] into a single benchmark package. CacheBench can be used to test the performance of the memory hierarchy. Each benchmark process includes reading cache, modifying and writing operations, and with one benchmark process at one time per VM. Transfer bandwidth is recorded.
- *Networking performance.* TCP/UDP performance is measured using Iperf.[4] The network bandwidth of VM's internal and cross-machine network connections are tested. For internal network connections,

[1] See MPBench at http://icl.cs.utk.edu/projects/llcbench/mpbench.html.
[2] See CacheBench1 at http://icl.cs.utk.edu/projects/llcbench/cachebench.html.
[3] See BLASBench at http://icl.cs.utk.edu/projects/llcbench/blasbench.html.
[4] See Iperf at https://code.google.com/p/iperf/.

the VM for sending data packets and VM for receiving data packets are located on the same physical machine. For the cross-machine scenario, the two VMs are provisioned on two different physical machines.

14.4.3 Test workflow

The general steps of virtual computing resource benchmarking (Figure 14.3) for each cloud platform include:

Step 1. Building an image with different virtual computing resource benchmark software packages installed for each open-source solution—To build an image, a VM needs to be launched with the configuration listed in Table 14.3 and can be further customized. The virtual computing resource benchmark software package should be installed on the basic VM. The installation and configuration of these benchmarkers, including UBench, Bonnie++, CacheBench, and Iperf, are described as follows.

- *CPU benchmarking.* The following commands show the installation of UBench on a Linux VM.

```
PROMPT>> wget ftp://ftp.uwsg.indiana.edu/pub/FreeBSD/
   ports/distfiles/ubench-0.32.tar.gz
PROMPT>> tar xvfz ubench-0.32.tar.gz
PROMPT>> cd ubench-0.32
PROMPT>>./configure
PROMPT>>patch -p0 < ubench-patch.txt
PROMPT>> make
```

The UBench software package has a small bug, and requires the use of a patch file `ubench-patch.txt` by running the command `patch -p0 < ubench-patch.txt` to fix the bug. Appendix 14.1 shows the content of the patch file.

- *I/O benchmarking.* The following commands are used to download and install Bonnie++.

```
PROMPT>> wget http://pkgs.repoforge.org/bonnie++/
   bonnie++-1.03e-1.el5.rf.x86_64.rpm
```

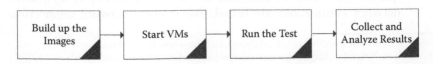

Figure 14.3 Virtual computing resource benchmark workflow.

```
PROMPT>> yum install gcc-c++ ## If g++ is not install
PROMPT>> rpm -i bonnie++-1.03e-1.el5.rf.x86_64.rpm
PROMPT>> /usr/sbin/bonnie++ ## Run Bonnie++ without
    options to check the usage
```

- *Memory benchmarking.* The CacheBench program can determine the performance parameters of a memory architecture subsystem. It is integrated into LLCbench. The following commands show the downloading of LLCbench and installation of CacheBench on a Linux VM.

```
PROMPT>> wget http://icl.cs.utk.edu/projects/
    llcbench/llcbench.tar.gz
PROMPT>> tar xvfz llcbench.tar.gz
PROMPT>> cd llcbench
PROMPT>> make linux-lam
PROMPT>> cd cachebench
PROMPT>> make
```

- *Networking benchmarking.* Iperf can be easily installed on any UNIX/Linux or Microsoft Windows system. The following commands show how to install the Iperf on Linux CentOS system.

```
PROMPT>> http://superb-dca2.dl.sourceforge.net/
    project/iperf/iperf/2.0.4%20source/iperf-
    2.0.4.tar.gz
PROMPT>> yum install gcc-c++
PROMPT>> tar xvfz iperf-2.0.4.tar.gz
PROMPT>> cd iperf-2.0.4
PROMPT>>./configure
PROMPT>> make && make install
```

After installing those benchmark tools, an image can be built from the VM for each solution.

Step 2. Starting different types of VMs—The five VM types (Table 14.3) can be launched from the images built in Step 1 on each solution.

Step 3. Running the test—This step requires creating the script to run each benchmarker sequentially. Typically, each benchmarker will be repeated three times, and with a wait for 60 seconds between each test. Details of the script for each benchmarker are as follows:

- *CPU benchmarking.* The scripts will repeat the UBench test three times with a wait of 60 seconds between each test.

```
testNum=3
i=0
while [ $i -lt $ testNum];
do
```

```
/Directory_for_ubench/ubench >> ubench.log ## run the
     test
sleep 60 ## wait for 60 seconds
let i=i+1
done
```

- *I/O benchmarking*. Bonnie++ benchmarking was done in a two-step fashion: (1) determining the smallest file size that invalidates the memory-based I/O cache, and (2) using a file size larger than the threshold size to test the real I/O performance (recommend using a file size that is two times the RAM size). File sizes range from 1024 KB to 40 GB.

  In this example, VM with 4 GB RAM size is used, and therefore the file size for writing to the hard drive should be 8 GB. To run the Bonnie++ benchmarking, the line /Directory_for_ ubench/ubench >> ubench.log should be changed to the following commands.

```
/usr/sbin/bonnie++ -d /mnt -s 8g -m centos_server
     -f -b -u root >> bonnie.log
```

  The actual benchmarking process for each time runs for minutes to hours depending on the file size to be written.

- *Memory benchmarking*. To run the CacheBench benchmarking, the line /Directory_for_ubench/ubench >> ubench.log should be changed to the following commands.

```
/Directory_for_ cachebench/cachebench -b -x1 -m9 -e1
     >> cachebench.log ## run the test
```

- *Networking benchmarking*. Iperf networking benchmarking is different from the previous two tests because two VMs are required for each test: one VM must be set as client and the other one as server. The server-side script to start an Iperf server listener process for the client connection is shown.

```
iperf -s -f m >& iperf.server.log.out &
```

  This command starts an Iperf listener on default port 5001. Iperf also needs to create a client process on the client VM to connect to the listener. In the experiment design, the Iperf test will repeat three times, and wait 60 seconds before the next round of tests for each VM type. Therefore, the client-side scripts to test the network performance between the server listener and client VM are as follows (server listener IP 199.26.254.162).

```
echo "Begin iperf test"
iperf -c 199.26.254.162 -P 4 -f m -w 256K -t 60 >>
    iperf.client.log.out
sleep 60 ## wait for 60 seconds
iperf -c 199.26.254.162 -P 4 -f m -w 256K -t 60 >>
    iperf.client.log.out
sleep 60
iperf -c 199.26.254.162 -P 4 -f m -w 256K -t 60 >>
    iperf.client.log.out
echo "Three iperf tests finished"
```

Step 4: Collect and analyze the results—When the benchmarking is completed, check the log files (e.g., ubench.log) for the CPU benchmarking result.

14.4.4 Test result analyses

- *CPU benchmarking results.* The final result is a score that reflects the CPU performances. Figure 14.4 shows a section of UBench output (Iperf.client.log.out file), including both CPU and RAM test scores. It can be observed that the VM gets a score of 2161708 for the CPU test.
- *I/O benchmarking results.* The default outputs are in comma-separated values (CSV) format including results of data reading and writing speed, number of seeks per second, and number of file metadata operations per second. The CPU usage statistics are also included in the benchmarking report for the following operations: (1) creating files in sequential order, (2) stating files in sequential order, (3) deleting files in sequential order, (4) creating files in random order, (5) stating files in random order, and (6) deleting files in random order. Figure 14.5 shows a sample benchmarking report.

  Results such as the "Sequential Output" in "Per Chr" methods are extracted from the reports and input into a table (Table 14.5).

```
Unix Benchmark Utility v.0.3
Copyright (C) July, 1999 PhysTech, Inc.
Author: Sergei Viznyuk <sv@phystech.com>
http://www.phystech.com/download/ubench.html
Linux 2.6.18-194.el5xen #1 SMP Fri Apr 2 15:34:40 EDT 2010 x86_64
Ubench CPU:   2161708
Ubench MEM:    340333
-------------------
Ubench AVG:   1251020
```

Figure 14.4 UBench output.

```
Version  1.03        ------Sequential Output------ --Sequential Input-
--Random-
                     -Per Chr- --Block-- -Rewrite- -Per Chr- --Block--
--Seeks--
Machine        Size K/sec %CP K/sec %CP K/sec %CP K/sec %CP K/sec %CP
/sec %CP
test           8G 15945  23 16881   3  7192   0 39739  34 42668   0
87.5   0
                     ------Sequential Create------ --------Random
Create--------
                     -Create-- --Read--- -Delete-- -Create-- --Read---
-Delete--
               files /sec %CP  /sec %CP  /sec %CP  /sec %CP  /sec %CP
/sec %CP
               16 21034  97 +++++ +++ +++++ +++ 15791  73 +++++ +++
+++++ +++
test,8G,15945,23,16881,3,7192,0,39739,34,42668,0,87.5,0,16,21034,97,
+++++,+++,+++++,+++,15791,73,+++++,+++,+++++,+++
```

Figure 14.5 A Bonnie++ benchmarking report sample.

Table 14.5 A Bonnie++ Test Result Sample on Eucalyptus Cloud Platform

		Seq. Output (KB/s)			Seq. Input (KB/s)		Rand. Input (seek/s)
		Char	Block	Rewrite	Char	Block	
Eucalyptus	Small	365	31561	16924	411	42135	91.9
	Medium	310	30155	16484	415	41901	135.5
	Large	312	30663	18457	458	45461	44
	Xlarge	308	27945	16440	477	40535	29.5
	XXLarge	308	29027	17025	378	41949	29.7

```
                  double RMW Cache Test

C Size          Nanosec         MB/sec          % Chnge
-------         -------         -------         -------
256             0.49            15473.64        1.00
384             0.46            16448.61        0.94
512             0.46            16559.79        0.99
```

Figure 14.6 CacheBench output.

The results indicate the I/O performance of the tested instance. The same benchmarking process is carried out on all platforms with different computing configurations. When all results are plotted, they reflect the performance differences between instances.

- *Memory benchmarking results.* CacheBench measures memory access time in nanoseconds and bandwidth in MB/sec. Figure 14.6 shows one test result in the output in the log file (cachebench.log).
- *Networking benchmarking results.* Iperf records the bandwidth in the unit of Mbits per second (Mbits/sec). Figure 14.7 shows the network

```
-----------------------------------------------------------------
Client connecting to 199.26.254.162, TCP port 5001
TCP window size: 0.25 MByte (WARNING: requested 0.25 MByte)
-----------------------------------------------------------------
[  4] local 199.26.254.163 port 34738 connected with 199.26.254.162 port 5001
[  5] local 199.26.254.163 port 34739 connected with 199.26.254.162 port 5001
[  3] local 199.26.254.163 port 34737 connected with 199.26.254.162 port 5001
[  6] local 199.26.254.163 port 34740 connected with 199.26.254.162 port 5001
[ ID] Interval        Transfer      Bandwidth
[  3]  0.0-30.0 sec    825 MBytes    230 Mbits/sec
[ ID] Interval        Transfer      Bandwidth
[  4]  0.0-30.0 sec    850 MBytes    238 Mbits/sec
[ ID] Interval        Transfer      Bandwidth
[  5]  0.0-30.0 sec    851 MBytes    238 Mbits/sec
[ ID] Interval        Transfer      Bandwidth
[  6]  0.0-30.0 sec    845 MBytes    236 Mbits/sec
[SUM]  0.0-30.0 sec   3370 MBytes    941 Mbits/sec
-----------------------------------------------------------------
```

Figure 14.7 Iperf output.

performance test result for the VM with IP 199.26.254.162 (IP of the client sending the request is 199.26.254.163).

14.5 TESTS OF GENERAL APPLICATIONS

14.5.1 Brief introduction of test aspects

This general application test includes the Java and database operation performance test (Table 14.4). SPECjvm2008 is a benchmark tool for the Java Runtime Environment (JRE) performance measurement. MySQL-Bench is used to test database operation performance using MySQL as an example.

14.5.2 Test design

- *JRE performance*—SPECjvm2008 contains several real-world applications and benchmarks for the purpose of evaluating Java functionality performance. SPECjvm2008 focuses on the performance benchmark of JRE in a single machine. In other words, it reflects the general performance of processor and memory while the I/O and network across machines have low dependence on SPECjvm2008 benchmarking. The metrics provided by SPECjvm2008 is operations per minute (ops/m), which includes:
  - *SPECjvm2008 base ops/m*—The overall performance from a fully complied base run.
  - *SPECjvm2008 peak ops/m*—The peak performance in the application run.
    This experiment includes installing and running the SPECjvm2008 on VMs launched by different cloud providers to benchmark the JRE performance on different clouds.

- *MySQL performance*—MySQL-Bench is used to test the performance of the MySQL database. It can test various types of database operations such as creating tables, inserting and deleting records, and querying on tables, which ensures a comprehensive test on the capabilities of the MySQL database supported by different solutions. The metrics provided by MySQL-Bench record the total time costs of executing all database operations defined by the benchmark tool in seconds.

14.5.3 Testing workflow

The detailed steps for producing final benchmarking reports for JRE and MySQL-Bench performance are similar to the virtual computing resource test (Figure 14.3):

Step 1. Build images with JRE benchmark software package or MySQL-Bench installed on each cloud platform.

- *Install and configure JRE.* To build a JRE benchmark image for different cloud platforms, a basic VM on each cloud platform should be started. JRE should be installed on the VM. The benchmarker SPECjvm2008 should also be downloaded and installed to the VM with the following command.

```
PROMPT>> wget http://spec.it.miami.edu/downloads/osg/
java/SPECjvm2008_1_01_setup.jar
PROMPT>> java -jar SPECjvm2008_1_01_setup.jar -i
console
```

- *Install and configure MySQL.* To build images with the MySQL benchmarker, VMs on each platform should be launched and checked for the version of MySQL. If an old version exists, remove the old version and install the latest version with the following commands.

```
PROMPT>> which mysql
PROMPT>> yum remove mysql-server
PROMPT>> yum remove mysql
PROMPT>> yum install mysql-server
PROMPT>> yum install mysql
PROMPT>> yum install mysql-devel
```

After MySQL is properly installed, MySQL-Bench, which is a benchmarking tool, can be installed with the following command.

```
PROMPT>> yum install mysql-bench
PROMPT>> cd /usr/share/sql-bench
```

Step 2: Start different types of VMs on different solutions—The five VM types (Table 14.3) can be launched from the images built in Step 1 on each solution.

Step 3: Run the test.

- *Test JRE performance.* To run the SPECjvm2008 test three times, the command line /Directory_for_ubench/ubench >> ubench.log should be changed to the following command.

```
/Directory_for_SPECjvm2008/run-specjvm.sh startup.
helloworld -ikv >> SPECjvm2008.log
```

- *Test MySQL performance.* To repeat a MySQL-Bench full test three times, the command line /Directory_for_ubench/ ubench >> ubench.log should be changed to the following command.

```
/usr/share/sql-bench/perl run-all-tests
    --user=ACCOUNT -password=MYROOTPASSWORD >> sql-
    bench.log
```

ACCOUNT and MYROOTPASSWORD should be replaced with the true user account and password for accessing the MySQL database.

Step 4: Collect and analyze the results—The log file (SPECjvm2008. log) should contain the three times test results for the JRE test. The log file sql-bench.log should contain the test results for the SQL-Bench test.

14.5.4 Test result analyses

- *JRE Performance.* The JRE test script generates a log file (SPECjvm2008.log) that records the number of operations in one second for three times and the detailed performance information. Figure 14.8 shows the output of a single test run. The result indicates that the VM performs at 141.51 operations per second.

SPECjvm2008 Base

n/a n/a
Oracle Corporation OpenJDK 64-Bit Server VM
Tested by: n/a
Test date: Wed Mar 20 21:34:55 EDT 2013

Benchmark	ops/m
startup	141.51

Figure 14.8 A sample test result of SPECjvm2008.

- *MySQL*. The log file (`sql-bench.log`) records time costs of executing all operations on the testing VM where the script resides. Figure 14.9 shows the output of a single test run captured in the log file. The first numerical value indicates the time cost in seconds. Therefore, the test costs 823 seconds on this particular machine.

14.6 CLOUD READINESS TEST FOR GEOSS CLEARINGHOUSE

14.6.1 Clearinghouse computing requirements

GEOSS Clearinghouse (CLH) (refer to Chapter 8 and Chapter 12), a geospatial portal with massive concurrent accesses, is used to test the concurrent performance of the cloud platforms.

14.6.2 Test design, workflow, and analysis

The matrix test conducted in Chapter 12 for the commercial cloud test could also be used in the open-source cloud platform test after revision. Figure 14.10 shows the test design on the three solutions: (1) CLH and JMeter are installed onto the three VMs with the open-source clouds and a traditional local server; (2) The four solutions (three cloud platforms plus one traditional computing platform) are connected through the Internet; (3) CLH can be tested by sending concurrent requests from one solution (client side) to the other three solutions (server side) with JMeter.

```
Start sql-bench:
TOTALS                          823.00  227.04    54.25  281.29 3425950
```

Figure 14.9 Sample test result.

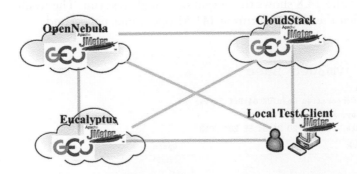

Figure 14.10 Matrix test of GEOSS Clearinghouse.

The three solutions' VMs can serve as servers, and each solution's VM will serve as a test client sequentially. Finally, concurrent performance results between these three solutions can be obtained from the test.

Chapter 12, Section 12.3 includes the detailed design, workflow, and test analysis about the CLH matrix, cloud load balance, and scalability tests.

14.7 CLOUD READINESS TEST FOR DUST STORM FORECASTING

14.7.1 Dust storm forecasting computing requirements

As introduced in Chapter 10 and Chapter 12, dust storm forecasting is both computational and communication intensive. Therefore, it is used to test the readiness of open-source solutions in support of computing intensive geoscience applications.

14.7.2 Test design

Chapter 12, Section 12.5 compared and tested the differences of EC2 and Nebula platforms, and benchmarked the hyperthreading capability. This section presents designs of several groups of experiments for benchmarking the open-source cloud solutions. The dust storm model is deployed on a local HPC cluster and three different open-source solutions.

Three groups of experiments are designed as follows:

1. *HPC versus clouds.* The virtual cluster built from VMs is compared to traditional clusters to quantify the overhead of transforming physical infrastructure into clouds. The result of this experiment would indicate how well the solutions support large-scale scientific computing. Within this experiment, different numbers of virtualized (from one to four VMs) and nonvirtualized computing resources are compared to investigate the impact of virtualized computing power, storage, and networking.
2. *Open-source solution comparison.* This experiment tests the capability of different solutions in supporting the computing- and communication-intensive applications with different numbers of VMs on the physical machines and three cloud solutions, respectively. The results indicate the performance of these cloud solutions for supporting scientific computing.
3. *Virtualization technology.* This experiment compares the performance of the same amount of computing resources virtualized using KVM and Xen.

14.7.3 Test workflow

The test workflow for each group is similar to Chapter 12, Section 12.5.3 (Figure 12.7): (1) building the images for each cloud platform; (2) writing the testing shell script for running the model and recording the performance; (3) starting the same number of VMs on each platform using the images; (4) configuring the model environments; (5) running the tests and collecting the results. Some specific configurations of the VMs for the three group tests designed in Section 14.7.2 are described as follows:

1. *HPC versus clouds.* The tests run the same model simulation on one, two, three, and four XXLarge VMs using different process numbers on the three solutions.
2. *Cloud middleware comparison.* The test adds another group of experiments with two large VMs on different solutions. Integrating the previous group performance results, the capabilities of different solutions for managing the VMs can be compared and tested.
3. *Virtualization technology.* This test runs the simulation on one, two, three, and four VMs that are created on the same solution (e.g., OpenNebula or Eucalyptus) but with two different popular open-source virtualization technologies KVM and Xen (Chapter 3, Section 3.4). The objective is to test the differences of KVM and Xen technologies.

14.7.4 Test result analyses

Section 12.5.4 introduces how to interpret the test result output of dust storm forecasting.

14.8 SUMMARY

This chapter introduces how to test open-source cloud computing solutions. Section 14.1 introduces the general background, community, and user feedback of three open-source solutions including Eucalyptus, CloudStack, and OpenNebula. Section 14.2 presents the test environments. Section 14.3 introduces the benchmarking procedures for the performance of managing cloud resources (e.g., VM) by the solutions. Section 14.4 introduces how to benchmark the virtual computing resource using different tools. This chapter also presents test strategies for general applications (Section 14.5) and geoscience applications, including the GEOSS Clearinghouse (Section 14.6) and the dust storm simulation (Section 14.7). The testing method, experiences, and scripts provided through the online Web site can also be used to test other open-source solutions. Huang et al. (2013) provides detailed test results of the three solutions.

14.9 PROBLEMS

1. What are the aspects that should be considered to test the virtual computing resources?
2. Please enumerate the tools that can be used to test the virtual computing resources.
3. What is the general workflow of testing the virtual cloud computing resources?
4. How do you test the open-source solutions with general applications?
5. What are the aspects that should be considered when testing the open-source solutions?
6. How do you test the capability of an open-source solution in supporting concurrent intensity?
7. How do you test the capability of an open-source solution in supporting computing intensity?
8. Read the results paper (Huang et al. 2013) and describe the results in 500 words. Discuss the dynamics of the results, that is, how the results may change.

APPENDIX 14.1 UBENCH PATCH FILE (UBENCH-PATCH.TXT)[1]

```
— - membench.c.old 2008-05-18 21:21:02.000000000 +0800
+++ membench.c 2008-05-18 21:38:20.000000000 +0800
@@ -23,6 +23,7 @@
#define MAX_CHILDS 128
#define MUFSIZE          1024
+#include <bits/time.h>
#include <sys/types.h>
#include <sys/times.h>
#include <stdio.h>
— - cpubench.c.old 2008-05-18 21:21:06.000000000 +0800
+++ cpubench.c 2008-05-18 21:38:16.000000000 +0800
@@ -22,6 +22,7 @@
#define CPUREFSCORE 50190
#define MAX_CHILDS 128

+#include <bits/time.h>
#include <sys/types.h>
#include <sys/times.h>
#include <stdio.h>
— - configure.old 2008-05-18 21:39:14.000000000 +0800
+++ configure 2008-05-18 21:39:42.000000000 +0800
```

---

[1] Note: Thank you to the anonymous contributor for publishing this patch online.

```
@@ -24,7 +24,7 @@
      i486)
cat <<! >> Makefile
CC = gcc
-CFLAGS = -O2 -m486 -Wall -malign-loops = 2 -malign-jumps =
    2 -malign-functions = 2 -fomit-frame-pointer
+CFLAGS = -O2 -m486 -Wall -falign-loops = 2 -falign-jumps =
    2 -falign-functions = 2 -fomit-frame-pointer
LDFLAGS = -s -lm
INCLUDES = -I.

@@ -33,7 +33,7 @@
      i586)
cat <<! >> Makefile
CC = gcc
-CFLAGS = -O2 -Wall -malign-loops = 2 -malign-jumps = 2
    -malign-functions = 2 -fomit-frame-pointer
+CFLAGS = -O2 -Wall -falign-loops = 2 -falign-jumps = 2
    -falign-functions = 2 -fomit-frame-pointer
LDFLAGS = -s -lm
INCLUDES = -I.

@@ -42,7 +42,7 @@
      i686)
cat <<! >> Makefile
CC = gcc
-CFLAGS = -O2 -Wall -malign-loops = 2 -malign-jumps = 2
    -malign-functions = 2 -fomit-frame-pointer
+CFLAGS = -O2 -Wall -falign-loops = 2 -falign-jumps = 2
    -falign-functions = 2 -fomit-frame-pointer
LDFLAGS = -s -lm
INCLUDES = -I.

@@ -51,7 +51,7 @@
      *)
cat <<! >> Makefile
CC = gcc
-CFLAGS = -O2 -Wall -malign-loops = 2 -malign-jumps = 2
    -malign-functions = 2
+CFLAGS = -O2 -Wall -falign-loops = 2 -falign-jumps = 2
    -falign-functions = 2
LDFLAGS = -s -lm
INCLUDES = -I.
```

REFERENCES

Barham, P., B. Dragovic, K. Fraser et al. 2003. Xen and the art of virtualization. *ACM SIGOPS Operating Systems Review* 37, no. 5: 164–177.

Huang, Q., C. Yang, K. Liu et al. 2013. Evaluating open-source cloud computing solutions for geosciences. *Computers & Geosciences* 59, 41–52.

KVM. 2010. Kernel-Based Virtual Machine. http://www.linuxkvm.org (accessed April 25, 2013).

Mucci, P. 2012. LLCbench (Low-Level Characterization Benchmarks). http://icl.cs.utk.edu/projects/llcbench/ (accessed April 28, 2013).

UBench. 2012. Ubench 0.32. http://www.tucows.com/preview/69604/Ubench (accessed April 25, 2013).

Chapter 15

GeoCloud initiative

Doug Nebert and Qunying Huang

The Geospatial Cloud Sandbox initiative (GeoCloud) was developed under the Geospatial Platform[1] activity to deploy and document geospatial cloud services based on shared Platform as a Service (PaaS) images that are codeveloped and rapidly configured by multiple agencies. This chapter introduces the methods and lessons learned from the project.

15.1 INTRODUCTION

With the flexibility, cost-effectiveness, and cost-efficiency of cloud computing, many federal agencies are adopting cloud technologies to cut costs and to make federal IT operations more efficient. In December 2011, U.S. Chief Information Officer (CIO) Vivek Kundra announced a "Cloud First" policy for federal agencies, requiring that all agencies move at least one system to a hosted environment.[2] The U.S. Department of the Treasury has moved Treasury.gov, SIGTARP.gov, MyMoney.gov, TIGTA.gov, and IRSOversightBoard.treasury.gov to the Amazon Elastic Cloud Computing (EC2) environment as part of the federal government's shift toward cloud services.[1]

GeoCloud is an annual prototyping initiative coordinated by the Federal Geographic Data Committee (FGDC), in collaboration with the U.S. General Services Administration (GSA) and the Department of Health and Human Services. It is designed as an incubator to test the feasibility of building and ultimately leading to promoting the concept of a GeoCloud community platform for agency applications within a secure cloud environment. With these standardized virtual geospatial servers, a variety of government geospatial applications can be quickly migrated to the cloud.

---

[1] See Geospatial Platform at http://www.geoplatform.gov/ (operated by the U.S. Federal Geographic Committee).

[2] See eWeek at http://www.eweek.com/c/a/Cloud-Computing/US-Treasury-Moves-Public -Web-Sites-to-Amazon-EC2-Cloud-201782/.

In addition, federal applications can achieve both infrastructure and platform savings:

- *Infrastructure savings*—The cost and labor for purchasing hardware, spending configurations and operations of the hardware, and implementing the scalability of applications can be significantly reduced. In addition, since these applications are hosted on the cloud, the costs for network and hosting infrastructure maintenance can be reduced.
- *Platform savings*—System-building time and effort are reduced by creating, maintaining, and sharing a common cloud-based community suite. In this way, agencies can achieve faster deployment and cost-effective development, and share system security profiles and documentation.

GeoCloud was designed to test and monitor externally hosted cloud data and service solutions for the geospatial domain and to support the inter-agency Geospatial Platform activities. In order to achieve this objective, multiple projects are nominated each year by federal agencies as existing public-facing geospatial data services. From the initial candidate projects, a set of common operating system and software requirements was identified as the baseline for the PaaS platform packages. Projects then are deployed using these common platform packages. In the process to develop best practices, cost and performance information is documented. Exemplar projects seek agency system security approval for continuing agency-sponsored operations in the cloud after the FGDC-sponsored hosting is completed. Two deployment environments (platforms) were defined for a set of projects, including open-source service stack on Linux 64 and a commercial service stack on Windows. Figure 15.1 shows the details about GeoCloud goals, activities, and outcomes.

15.2 GeoCloud ARCHITECTURE

The GeoCloud architectural framework includes three layers (Figure 15.2): (1) the infrastructure layer, (2) the PaaS layer, and (3) the application layer. The infrastructure layer integrates the cloud computing resources (e.g., virtual machine [VM], storage, and networking) and basic operation system suites for managing the VMs. The PaaS layer is the focus of the GeoCloud deployments and includes the common geospatial service software to enable user interface customization by the projects (applications). In this regard, the PaaS solutions proposed include tools for programming and customizing user interface components. This is very different from the typical Software as a Service (SaaS) environment where the interfaces have limited customization and extension potential.

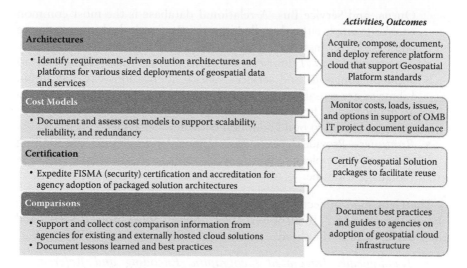

Architectures	**Activities, Outcomes**
• Identify requirements-driven solution architectures and platforms for various sized deployments of geospatial data and services	Acquire, compose, document, and deploy reference platform cloud that support Geospatial Platform standards
Cost Models	
• Document and assess cost models to support scalability, reliability, and redundancy	Monitor costs, loads, issues, and options in support of OMB IT project document guidance
Certification	
• Expedite FISMA (security) certification and accreditation for agency adoption of packaged solution architectures	Certify Geospatial Solution packages to facilitate reuse
Comparisons	
• Support and collect cost comparison information from agencies for existing and externally hosted cloud solutions • Document lessons learned and best practices	Document best practices and guides to agencies on adoption of geospatial cloud infrastructure

Figure 15.1 (See color insert.) GeoCloud Goals, Activities, and Outcomes.

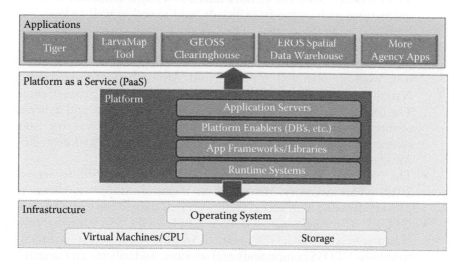

Figure 15.2 (See color insert.) GeoCloud architectural framework.

The PaaS layer consists of the following services:

- *Application Servers*—Servers provide the deployment environment for actual business applications with access to platform enablers, frameworks, and runtimes, including programming and user interfaces and payload format standards endorsed by the FGDC.
- *Platform Enablers*—Enablers provide core-supporting capabilities for developing, testing, and deploying code, including DBMS, Directory,

Queue, and Service Bus. A relational database is the most common enabler example either bundled into the PaaS or linked from a remote database server instance.

- *Frameworks and Libraries*—Frameworks provide API access to common functions and services, which applications can rely upon. Libraries are reusable code modules that can be called directly from an application. They can reduce time and expense, and free developers from having to build common code and behaviors.
- *Runtimes*—Provide the execution support for developing and running the code. Examples include Java, Python, and Microsoft Common Language Runtime.

The GeoCloud projects act as a surrogate for the application layer, as shown in Figure 15.2. In 2011 and 2012, the following projects completed their deployment in the cloud:

- *Topologically Integrated Geographic Encoding and Referencing (TIGER)/Line* (U.S. Census Bureau 2010)—TIGER/Line are spatial data extracted from the Census Bureau's TIGER database, containing features such as roads, railroads, and rivers, as well as legal and statistical geographic areas. The Census Bureau currently offers the datasets to the public for downloading from a census data server.
- *LarvaMap tool*[1]—This is a tool for modeling trajectories of larval fish, used in National Oceanic and Atmospheric/Administration National Marine Fisheries Service/Alaska Fisheries Science Center. The tool is intended to be used by resource managers and fisheries scientists to model the dispersion of fish larvae under varying environmental conditions. The tool is developed with Java and hosted by Apache Tomcat, and the OS of the tool is Windows.
- The National Wetlands Inventory (U.S. Fish and Wildlife Service 2012)—This inventory, along with its Wetlands Mapper application, provide data and map access to all wetlands data in the United States.
- *GEOSS Clearinghouse* (GEOSS 2011)—GEOSS Clearinghouse (Chapter 8) is a FGDC, GEO, and NASA project that connects directly to various GEOSS components and services, and collects and searches the distributed data and services via interoperable mechanisms.

15.3 GeoCloud ACTIVITIES

Figure 15.3 shows the GeoCloud primary activity cycle including: (1) creating prototype platforms, (2) validating with agency applications, and (3) documenting and promulgating.

---

[1] See NOAA Fisheries at http://www.afsc.noaa.gov/databases.htm.

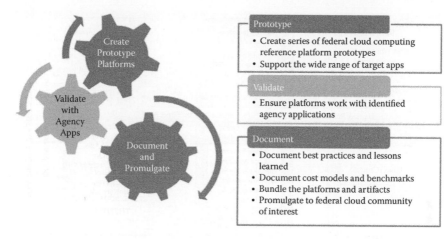

Figure 15.3 GeoCloud primary activity cycle.

Table 15.1 Application and Basic Software Package[a]

Applications	Basic Software Package
National Wetlands Inventory (NWI) Wetlands Mapper	AWS, Windows 2008, ArcGIS Server
U.S. Census TIGER/Line Downloads	AWS, Linux 64 (CentOS)
Integrated Ocean Observing System Catalog and Viewer (NOAA)	AWS, Linux 64 (CentOS), GeoNetwork
NOAA ERDDAP	AWS, Linux 64 (CentOS), THREDDS
EPA Lakes and Ponds	AWS, Windows 2008, ArcGIS Server
Particles in the Cloud (NOAA) particle tracking computational service for air or water dispersion/diffusion	AWS, Linux 64 (CentOS)
GEOSS Clearinghouse	AWS, Linux 64 (CentOS)
USDA FSA or NRCS data service application	AWS, Windows 2008, ArcGIS Server

[a] See Federal Geographic Data Committee GeoCloud Platform at http://semanticcommunity.info/@ api/deki/files/10504/GeoCloud_Platform_Business_Use_Cases.pdf.

15.3.1 Creating prototype platforms

With the popularity and reliability of Amazon Elastic Cloud Computing (EC2), it was selected as the primary public cloud computing environment for various sizes and numbers of VMs. Software requirements for the candidate platforms were dictated by the projects nominated by federal agencies. Standards are dictated by the FGDC-endorsed standards list, and commonalities were identified in defining the stack. Table 15.1 shows the basic software required for each project.

After analyzing the basic software requirements and the dependent packages for each project, GeoCloud shared platforms for different

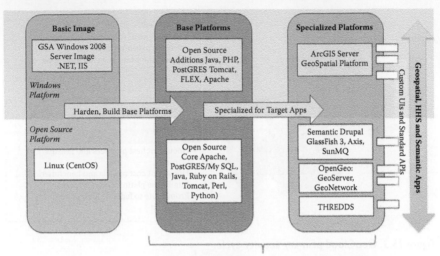

Tiers (e.g., database, app server) can be split or combined as needed

Figure 15.4 (See color insert.) GeoCloud platform creation and customization for federal geospatial applications. (From Nebert, 2010, www.fgdc.gov/ngac/meetings/december-2010/geocloud-briefing.pptx.)

federal geospatial applications were designed and deployed as shown in Figure 15.4. The basic images are of two types of OS—Linux (CentOS 5.5) and the Windows 2008 server. The Windows images are directly from Microsoft with a license passing through in the form of additional per-hour fees. After the basic image has been built from scratch, the most common software packages are installed on both Linux and Window images as base platforms, such as Java, Apache Tomcat, and MySQL.

Nationally, the base platforms could be further customized as specialized platforms for different domain applications. Those specialized platforms are equipped with different geospatial open-source packages, and they are divided into four categories according to the requirements of these identified applications: (1) ArcGIS Server, (2) Semantic Drupal, (3) OpenGeo, and (4) THREDDS.

In fiscal year 2012, two standard PaaS images were developed to support a standardized deployment including a limited application layer since the application servers include User Interface (UI) and Application Programming Interface (API) customization support. The selection of only two geospatial service suites, ArcGIS Server and OpenGeo Enterprise Suite, was based on the consolidated application requirements of the federal agencies. The support of only two PaaS images (platform solutions) allows the GeoCloud activity to coordinate and share security approaches, software packaging and updates, and develop a well-documented hosting environment for multiple agencies and application domains. This further simplifies the certification and accreditation of all deployed virtual "systems" that use these platform images as they are all clones of a well-known environment.

15.3.2 Validate with agency applications

After creating base and specialized platforms (Figure 15.4), selected applications were able to be deployed and tested on those platforms for validation. For example, the National Wetlands Inventory (NWI) Wetlands Mapper was deployed to the ArcGIS Server platform, whereas TIGER/Line and GEOSS Clearinghouse were directly migrated to the base platforms. During this process, the base and specialized platforms were adjusted and customized based on the test results of selected applications.

15.3.3 Document and promulgate

In order to provide the guidance for the federal agencies when seeking cloud solutions, GeoCloud provides a variety of documents including:

- Best practices and lessons learned.
- Cost models and benchmarks.
- Bundled platforms, updates, and artifacts.

All those documents are available on the FGDC Web site[1] and the GeoCloud community portal[2] as well. The project teams involved in the GeoCloud initiative have been working on promulgating to federal cloud communities of interest through a variety of activities:

- Publishing the deployment and test results on the international conference (Huang et al. 2010).
- Presenting the GeoCloud activities at professional and community conferences, such as the Federation of Earth Science Information Partners (ESIP), the Association of American Geographers (AAG), and the 2012 Esri International User Conference.

One significant outcome of the GeoCloud documentation and promulgation phase is the development of the common PaaS images and architectural pattern—reusable software image that mounts all configuration and data from an attached, persistent EBS volume. As mentioned previously, the focus on only two reusable PaaS solutions greatly simplifies the development,

---

[1] See FGDC at http://www.fgdc.gov/initiatives/geoplatform/geocloud.
[2] See Cloud Community Platform Portal at http://geocloud.eglobaltech.com/.

maintenance, and security profiling of all systems that use them for deployment across the government. Thus, the images and the scripts used to provision them are key resources of benefit from the GeoCloud initiative, geospatially enabling the Obama Administration's Shared Services agenda.

15.4 GeoCloud SECURITY

There are two levels of security concerns for the GeoCloud projects: the security of cloud platform Amazon EC2 used to host different government applications, and the GeoCloud security operations for administering the platforms and applications within the cloud.

15.4.1 Amazon Web Services (AWS) Security

Security and data privacy are primary concerns of system owners and may hamper agency adoption of cloud solutions. After conducting an in-depth investigation about current popular cloud vendors in terms of the capabilities and security strategies, Amazon EC2 was selected as the platform for GeoCloud. AWS has a number of certifications, including ISO 27001, and level 1 certification for financial security and the passage of personal identification information (PII) during the period of GeoCloud initiatives (AWS Security 2013). It has garnered approval for hosting 12 public-facing federal systems including Recovery.gov and Treasury.gov based on a Federal Information Security Management Act (FISMA) Low Accreditation, AWS has internally completed FISMA–Moderate level certifications, documented procedures, and gap analysis as of early 2013.

Amazon EC2 IaaS predates the Federal Risk and Authorization Management Program (FedRAMP) certification process but Amazon has provided documentation and security controls as part of agency-initiated FedRAMP approvals. It is expected that AWS will achieve FedRAMP approval such that it may be readily leveraged by any federal agency.

Although AWS offers its infrastructure to the cloud consumers, it does not provide a physical private cloud. However, virtual private clouds, such as GovCloud,[1] accessed through virtual private networks (VPNs), are available.

In May 2013, AWS received FedRAMP security approval for its infrastructure offerings, submitted by the Department of Health and Human Services (HHS) for use within its IT modernization activities. This Authority to Operate (ATO) can be leveraged by other federal agencies to further expedite the (virtual) systems approval process (see the press release at http://phx.corporate-ir.net/phoenix.zhtml?c= 176060&p=irol-newsArticle&ID=1822454&highlight).

---

[1] See AWS at http://aws.amazon.com/govcloud-us/.

15.4.2 GeoCloud security operation

GeoCloud has a separate product credential for each project to log in through the AWS Web management console. With this credential, the cloud consumer (who is the system administrator of each project) is able to only access and manage cloud services dedicated for the project through the AWS Web management console. The credentials were generated and provided by the GeoCloud system coordinator. This can achieve project level privacy so that each system administrator will not be able to impact or crash the data and applications of other projects.

The product-level credential enables the cloud consumers (system administrators for projects) to provision and release the VMs, and to configure the network access for the targeted applications. Several high-level operations are not authorized for project level system administrators, such as creating a new shared OS/software image based on the running instance. Since the cloud images for federal applications must comply with several certifications, authorizations, and regulations, only selected GeoCloud system coordinators are able to customize and build shared images.

15.5 OPERATIONAL COST IN THE CLOUD

The cost evaluation for each of the initial projects was performed using an online calculator and is based on data transfer, storage, CPU, and demand requirements. Most projects are feasibly hosted in AWS (~$350–$500/ month). The monthly cost spent on EC2 cloud from July to October 2010 for TIGER/Line and GEOSS Clearinghouse are shown in Figure 15.5. It can be observed that the monthly cost for the TIGER/Line project is around $300. The average monthly cost for GEOSS Clearinghouse is about $270.

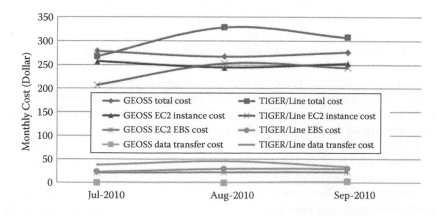

Figure 15.5 The monthly costs for TIGER/Line and the GEOSS Clearinghouse project.

However, some projects were cost-prohibitive in the cloud due to large data storage or transfer costs. For example, hosting all of USGS EROS Spatial Data Warehouse in a cloud environment could potentially be very expensive. The services identified in the initial year effort would require six large Windows instances, 5 TB of EBS storage, 50 GB input per month, and 500 GB output with no S3, or elastic LBs. These collectively would cost anywhere from $2,500 to $4,000 per month (SDW 2010).

15.6 DISCUSSION

After one year of operation and monitoring of selected projects in the cloud in 2010–11, lessons learned have been recorded for federal agencies' future adopting cloud solutions.

Security approval process. A big challenge for GeoCloud was verifying agency system security approvals based on documentation from Amazon. FISMA requires federal agencies to develop, document, and implement an information security system for its data and infrastructure. To migrate the GeoCloud federal applications onto EC2, each project had to perform similar level risk assessment analysis as FISMA. For example, the TIGER/Line project was able to attain FISMA level assurance due to the conclusions reached in the risk assessment that was conducted by the Office of Information Security (OIS). The OIS risk assessment focused on three primary security elements: (1) confidentiality, (2) integrity, and (3) availability. TIGER/Line data have a very low confidentiality impact level because of their public nature. Furthermore, it was determined that reconstitution of the TIGER/Line system internally was extremely fast and efficient resulting in effective compensating controls to prevent any attack on the availability of the system within AWS. Based on the above analysis coupled with the low probability of attack on this type of public information resulted in the conclusion that the overall risk associated with operating this system within AWS was very low (TIGER/Line 2010).

Software deployment. Currently, all GeoCloud projects share one of two common platform images including the primary application server, supporting libraries, and Web deployment tools including MySQL, PostgreSQL, gcc, Java, and Tomcat. However, the storage and performance of the system may be affected since more software than required has been installed and running. Configuration scripts using Amazon CloudFormation have since been developed to select and activate specific software.

Time-to-deploy. Based on the experiences of the system administrators for those federal applications who were novel cloud users, it takes around (1) one to two weeks to get familiar with the AWS services, (2) two days to customize the basic system running from a public machine image or hardened image, (3) two days to script the processing of customizing the applications, and (4) one week to explore the load balancer, auto-scaling capabilities, elastic IP, data and code backup, and recovery. Since becoming familiar with the AWS takes a long time for a nonexpert, it might be better to have an introduction workshop for those new to the AWS to get a quick start. This book could serve part of that requirement.

Failover, redundancy. AWS provided the capability to expand system resources as needed. According to the configuration, additional CPU and memory could be made available if required (TIGER/Line 2010). However, when the concurrent number of users exceeds 600, there may be some response failures (GEOSS Clearinghouse 2011). Multiple copies can be distributed to different cloud regions for further protection against a system crash, offering rapid failover capabilities. Using CloudFormation scripts, data volumes can be backed up using the snapshot process on a frequency determined by the publisher.

15.7 SUMMARY

Using the GeoCloud initiative as an example, this chapter illustrates how to migrate federal applications onto the cloud. GeoCloud project also indicates how to build up a geospatial cloud community platform for different applications. This chapter briefs the GeoCloud background (Section 15.1), architecture (Section 15.2), activities (Section 15.3), security (Section 15.4), cost (Section 15.5), and lessons learned (Section 15.6) of multiple federal applications across multiple agencies for building a geospatial community platform.

15.8 PROBLEMS

1. What is the GeoCloud initiative?
2. What are the benefits of building a common geoplatform?
3. What is the GeoCloud architecture?
4. Enumerate the GeoCloud activities.
5. What are the security considerations? How is security achieved for federal applications?
6. What is the operational cost in the cloud for a typical Web application? For data-intensive applications?
7. What are the lessons learned?

REFERENCES

AWS Security. 2010. http://aws.amazon.com/security/ (accessed March 13, 2013).
GEOSS. 2011. GEOSS Clearinghouse Report. http://www.fgdc.gov/initiatives/geoplatform/geocloud/reports/fgdc-geocloud-project-report-geonetwork.pdf (accessed March 13, 2013).

Huang, Q., C. Yang, D. Nebert, K. Liu, and H. Wu. 2010. Cloud Computing for Geosciences: Deployment of GEOSS Clearinghouse on Amazon's EC2. In *Proceedings of ACM SIGSPATIAL International Workshop on High Performance and Distributed Geographic Information Systems (HPDGIS)*, November 2, San Jose, CA.

Nebert, D. 2010. Deploying Federal Geospatial Services in the Cloud. www.fgdc.gov/ngac/meetings/december-2010/geocloud-briefing.pptx (accessed March 13, 2013).

SDW. 2010. EROS Spatial Data Warehouse Report. http://www.fgdc.gov/initiatives/geoplatform/geocloud/reports/fgdc-geocloud-project-report-eros-sdw.pdf (accessed March 13, 2013).

U.S. Census Bureau. 2010. Topologically Integrated Geographic Encoding and Referencing (TIGER) TIGER/Line, 2010 Final Report. ftp://ftp2.census.gov/geo/tiger/TIGER2010.

U.S. Fish and Wildlife Service. 2012. National Wetlands Inventory Database Metadata. http://www.fws.gov/wetlands/data/metadata/FWS_Wetlands.xml (accessed March 13, 2013).

Part V

Future directions

This part reviews and then discusses the future research and development directions for spatial cloud computing. Chapter 16 introduces the intensities of data; computing, concurrent, and spatiotemporal; and discusses potential solutions and the research remaining to address the intensity challenges using cloud computing. Chapter 17 introduces future research needs from the three aspects of geoscience vision, technical advancements, and synergistic advancements between cloud computing and social sciences.

Chapter 16

Handling intensities of data, computation, concurrent access, and spatiotemporal patterns

Qunying Huang, Zhenlong Li, Kai Liu, Jizhe Xia, Yunfeng Jiang, Chen Xu, and Chaowei Yang

The advancements of geoscience involve dealing with intensities of data, computation, concurrent access, and spatiotemporal patterns. This chapter discusses each of the intensities, and introduces through examples the potential solutions to address the intensity challenges using cloud services.

16.1 INTRODUCTION

The Earth and its belongings evolve in a four dimensional (4D) world. Geosciences study the patterns and possible trends of this evolution of specific or integrated phenomena, such as dust storms and climate change. The grand challenges facing us in the 21st century are inevitably related to this 4D evolution, and addressing the challenges that require us to understand the principles and relationship behind the evolution and to predict potential future evolution based on different factors for different user communities. For example, we need to integrate the domains of land, ocean, and atmosphere processes to better understand how the climate is changing. This human knowledge pursuant to the application process involves large amounts of data (e.g., for climate change), significant computing requirements to process increasing volumes of data and complex algorithms (e.g., for social media processing), and large numbers of simultaneous connections to the knowledge and computing (e.g., for emergency response to tsunamis and earthquakes). The three intensities of data, computation, and concurrent access are all based on spatiotemporal intensity, which we have to deal with on a variety of scales and resolutions. These intensities naturally match the characteristics of cloud services, such as scalable computing resources for concurrent access, virtually unlimited computing power for computing intensity, and global, available data centers by cloud providers for distributed locations of data (Yang et al. 2011a). This chapter reviews and discusses the future research directions of using cloud services to address these intensities.

16.2 BIG DATA

16.2.1 Introduction

Big data generally refer to four *Vs*: volume, velocity, variety, and veracity.[1] Volume refers to the size of data; velocity indicates that big data are sensitive to time; variety means big data comprise various types of data with complicated relationships, and veracity indicates the trustworthiness of the data. In the geospatial domain, big data are mainly produced from (1) the various sensors with different spatial, temporal, and spectral resolutions that record physical phenomena, (2) the various scientific models simulating and predicting the physical phenomena, and (3) the exponential growth of the relationship among big data. How to effectively store and process big data for efficient data access and analysis poses critical challenges:

- *Big data management challenge*—Effective management of big data for data discovery, processing, and visualization is a fundamental and yet critical premise for handling data intensities. The intrinsic heterogeneity of big data due to diverse data collecting methods and usage scenarios poses grand challenges for data searching, integration, and management (Li et al. 2011). The large volume of data makes big data management even more complicated. Storing big data requires highly scalable storage devices that are able to easily scale up to fulfill the request, and easily scale down to minimize the cost.
- *Big data processing challenge*—Under most circumstances, managing big data is not the final purpose. The ultimate goal is to transform these data into information, knowledge, and insights through processing and analyzing. However, processing big data within an acceptable time frame poses a critical challenge. For example, 10 terabytes are normal data volume produced by the climate model in Climate@ Home (Chapter 9). Suppose that the read/write speed for the storage device is 50 megabytes per second, reading 10 terabytes of data on one machine needs 2.5 days, reading 100 terabytes of data requires almost a month.
- *Big data visualization challenge*—Visualizing big data is essential for understanding geographic phenomena by converting data and information into graphical representations. Based on the visualizing dimension, big geospatial data visualization can be divided into two categories: two-dimension (2D) visualization and multiple-dimension (multi-D) visualization, such as 3D and 4D (3D with time). 2D visualization is relatively straightforward and the techniques are more mature, such as pixel-based visualization for intuitive viewing and

---

[1] See IBM at http://www-01.ibm.com/software/data/bigdata/.

various diagrams for visual analytics. However, multi-D visualization still poses challenges due to the characteristics of multiple dimensions and variables as well as the complexity and volume of the spatiotemporal data (Yang et al. 2013).

16.2.2 An example with Climate@Home

Climate@Home utilizes computing resources contributed by citizens to enable massive numbers of climate model runs (ModelE) in a distributed environment. ModelE may run many times on thousands of volunteer machines around the world. Each model run with a 10-year simulation will generate 10 gigabytes of data in both 3D and 4D containing 566 climatic variables. The one-hundred model runs with a 200-year simulation that will output 20 terabytes of data (Li et al. 2013). Testing and optimizing climate prediction using ModelE requires thousands of model runs, which could generate petabytes of data. These data are big in that the volume is massive, and the accumulating speed is fast.

16.2.3 Solutions

Three solutions are used to address the challenges, respectively.

* *Metadata, spatiotemporal index, and a distributed file system (DFS)*—
  Metadata is structured descriptive information about the data and has been widely used in diverse data-intensive applications (Singh et al. 2003; Yee et al. 2003). Metadata is important in big data management for (1) supporting massive data organization by describing the design and specification of the complex data structures, (2) enabling efficient data discovery and access by describing the data content, and (3) serving as the fundamental component for heterogeneous data integration.

  While metadata serve as a mechanism to help describe and manage big data, the performance of data query and access is critical for on-demand geospatial applications. The indexing mechanism has the potential to significantly improve data discovery by providing efficient data retrieval algorithms (Theodoridis et al. 1996). The spatiotemporal index integrates the spatiotemporal principles (Yang et al. 2011b) and indexing techniques to provide high-performance search capabilities for data-intensive geospatial applications.

  A DFS (Chapter 3, Section 3.4.1) is a potential solution for the big data storage challenge within a cloud service. Because DFS provides the capability for transparent replication and fault tolerance, the reliability is enhanced. Furthermore, DFS is designed to utilize the commodity hardware (e.g., PCs) in a loosely coupled environment; therefore, the storage capacity can be easily scaled up and scaled down.

- *Parallel computing with cloud services*—With respect to handling the big data processing challenge, the integration of parallel computing and cloud computing is utilized (Fox et al. 2010). Specifically, MapReduce (Lämmel 2008), a parallel data processing framework pioneered by Google, is leveraged, and Eucalyptus (Chapter 13) is selected as the cloud computing platform. MapReduce is designed to process big data in parallel by utilizing commodity PCs as computing nodes. Typically, MapReduce is based on DFS. The integration of DFS and MapReduce provides a reliable, scalable, and effective solution for storing and processing big data. Hadoop MapReduce and Hadoop Distributed File System (HDFS) (Chapter 3, Section 3.4.3) open-source implementations of MapReduce and DFS, are employed to implement the data storage and processing framework. In addition, Eucalyptus is utilized to build a private cloud service with flexible computing capability. The VMs launched in this private cloud can be used as the storage and processing nodes to further improve the scalability as well as reduce the cost.

- *Graphics Processing Unit (GPU)-based distributed visualization in cloud services*—Visualizing geospatial data in multi-D can produce more intuitive and visual effects and thus help users effectively explore the potential patterns in scientific data. The traditional visualization techniques with a regular visualization pipe cannot fulfill the performance requirement when visualizing big volume multi-D data, especially when the time dimension is taken into account. GPU-based visualization is leveraged to tackle this challenge by leveraging the graphics hardware acceleration as well as parallel computing. For example, the CUDA platform[1] published by NVIDIA provides a powerful GPU parallel computing capability to enable faster rendering speed. In addition, since the data volume is large, a distributed visualization strategy is adopted to further improve the visualizing performance by dividing the large dataset into small volumes and rendering them in a parallel manner. The GPU-enabled cloud resources (e.g., Amazon EC2 GPU instances) enable us to use and implement the GPU-based visualization model flexibly (see Chapter 17, Section 17.2.5).

16.2.4 Remaining problems and future research

The potential approaches provide a comprehensive solution for handling the data intensity from big data management; big data processing to big data visualization. However, the following three aspects still require further studies.

---

[1] See NVIDIA at http://www.nvidia.com/object/nvidia-iray.html.

- *Storage and computing capacities* are a big concern in dealing with data-intensive applications. For the storage capacity, the cloud storage services such as Amazon S3 and Elastic Block Store (EBS) could be leveraged, however, the cost for hosting and transferring the data in a commercial cloud storage service may be expensive given the size of the data. The same situation also applies to the computing capacity. We can establish a private cloud service to reduce the cost, but this is not always the case. The trade-off between using commercial services and establishing a private cloud service needs to be carefully examined based on the organization's existing IT infrastructure. Given the ever-growing data volume and the limited (budget concerns) storage and computing capacity, further studies are needed on (1) how to eliminate the redundant and useless data and information through smart data preprocessing/cleaning techniques, (2) how to reduce the data size through effective data compressing algorithms, and (3) how to accurately calculate the storage and computing provisions when using commercial cloud services.

- *On-demand big data access* is critical for real-time geospatial applications, particularly for those in a Web-based environment. For most applications, when data size is small, on-demand data access is not an issue; when dealing with big data, things are different. The parallel computing technique has great advancements in batch processing large volumes of data, which is suitable for systems where the latency is tolerable. However, real-time querying and accessing of terabytes of data is still a challenging problem. Further studies are required on (1) how to develop more efficient spatiotemporal indexing techniques for data searching, and (2) how to optimize the parallel computing technique for data subset and process.

- *The gap between MapReduce and scientific data*—MapReduce excels at processing unstructured data such as texts, documents, and Web pages. However, most geoscience applications deal with structured data such as relational databases and semistructured data such as array-based scientific data. Generally, three approaches are commonly used for digesting science datasets (e.g., NetCDF) in Hadoop MapReduce: (1) binary to text, which transforms the binary-based dataset into a text-based dataset (Zhao et al. 2010); (2) data reorganization, which reorganizes the original dataset without transforming it into another data format, and stores it in Hadoop-supported files, such as Sequence Files[1] (Duffy et al. 2012); and (3) developing a middleware on top of Hadoop for handling the scientific data (Buck et al. 2011), such as SciHadoop (Buck et al. 2011). These approaches to some extent fill the gap between Hadoop and geoscience data, and

---

[1] See the Hadoop Wiki at http://wiki.apache.org/hadoop/SequenceFile.

provide a general guideline for leveraging Hadoop in handling big data challenges in the geospatial domain. Nevertheless, further studies are necessary for improving the completeness and performance of these approaches as well as investigating other possible solutions.

16.3 COMPUTING INTENSITY

16.3.1 Introduction

Computing intensity is another issue that needs to be addressed in geosciences. In the context of geoscience, computing-intensive issues are normally raised by data mining for information/knowledge, parameters extraction, and phenomena simulation (Yang et al. 2011a).

- *Data mining for information/knowledge*—Many data-mining technologies have been investigated to better understand whether observed time series and spatial patterns are within the Earth subsystems, including biosphere, atmosphere, lithosphere, and social and economic systems. Interactions among those subsystems within spatiotemporal dimensions are intrinsically complex and data-mining algorithms and processes to explore such interactions are computing intensive (Donner et al. 2009). For example, exploring the interaction between the global carbon cycle and climate system are computing intensive (Kumar et al. 2006).
- *Parameter extraction*—Complex geophysical algorithms are utilized to obtain phenomena parameters (e.g., temperature) from massive Earth observation data. However, the execution of complex algorithmic processes to extract the parameters are extremely computing intensive. For example, the computation and storage requirements for deriving regional and global water, energy, and carbon conditions from multisensor and multitemporal datasets far exceed the capability of a single workstation (Kumar et al. 2006).
- *Phenomena simulation*—Simulating geospatial phenomena is especially complex when considering the full dynamics of the Earth system phenomena. An example is modeling and predicting cyclic processes (Donner et al. 2009), including ocean tides (Cartwright 2000), earthquakes (Schuster 1897), and dust storms (Chapter 10).

16.3.2 An example with digital evaluation model interpolation

Digital Evaluation Model (DEM) refers to the digital representation of ground surface topography or terrain, and is also widely known as a Digital Terrain Model (DTM). A DEM can be represented as a raster (a grid of cells) or vector (a triangular irregular network) in a Geographical Information System (GIS) (Audenino et al. 2001). The interpolation of DEMs for large geographic areas could encounter challenges in practical applications, especially for Web applications such as terrain visualization, which requires a fast response. In addition, the demand of computing exceeds the capacity of a traditional single processing unit that is only able to conduct serial processing (Huang and Yang 2011) (see Chapter 5, Section 5.3).

16.3.3 Solutions

Enhancing computing power, which contains CPU, memory, and disk is the basic solution to solve the computing issues from a hardware aspect. Multicore and many-core central processing units (CPUs) or GPUs, memory with large capacity and cache, and disks with high speed I/O are widely used for the deployment of computing applications. However, common approaches to address the computing problems is to use parallel programming where computing tasks are divided and executed on multiple CPU or GPU cores on a single machine or multiple machines (Huang et al. 2012).

16.3.3.1 Cloud CPU computing

The computing speed of the traditional serial-based computing model on a single machine cannot keep up with increasing computing demands. High Performance Computing (HPC) or grid computing (Armstrong et al. 2005; Huang and Yang 2011) have been used to address increasingly larger computing issues in geographic science problems, such as dust storm simulation. Such a large computing intensive problem can be quickly solved by dividing it into many subproblems, and leveraging multiple computational resources to address those subproblems in parallel. Taking DEM interpolation as an example, the DEM domain can be divided into several equal-sized subdomains, and those subdomains can be processed by different computing resources.

Figure 16.1 shows the performance of DEM interpolation with grid computing where traditional distributed CPU cores are used to handle the calculation (Huang and Yang 2011). However, it is not possible for every organization or terminal user to be equipped with HPC infrastructure. The deficiency of resources has hampered the advancements of HPC. Fortunately, cloud computing offers a powerful and affordable alternative to run large-scale computing intensive tasks (Huang et al. 2013). Chapter 5, Section 5.3 introduced the detailed steps to run DEM interpolation in EC2.

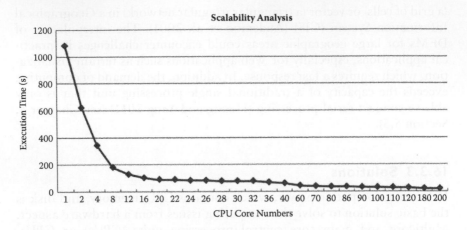

Figure 16.1 DEM interpolation with CPU parallel processing.

16.3.3.2 Cloud GPU computing

The emergence of GPU computing in recent decades provides scientists and engineers more optional computing techniques to solve complex science, engineering, and business problems. Compared to multicore CPU computing, many-core GPU computing has been proved a better choice in many cases, especially in improving rendering efficiency on geospatial visualization (Li et al. 2012).

Figure 16.2 shows the performance of GPU compared to CPU in supporting the DEM interpolation. The interpolation is executed separately on a CPU or GPU device with the same interpolation algorithm upon which each cell value is calculated based on the elevation value of its neighbors. With CPU computing, multithreads are used to improve the performance. With GPU computing, CUDA libraries, which support parallel computing, are used. Each grid cell in DEM computes on one GPU thread. While using GPU computing, the interpolation can only be executed in GPU memory rather than in CPU memory. Therefore, it is necessary to transfer DEM data from CPU memory to GPU memory. After finishing the processing, the resulting output could be shifted back to CPU memory.

GPU as a novel computing technique has been gradually integrated into the cloud computing framework, which is called *Cloud GPU*. Presently, consumers can easily create GPU virtual machines (VMs) to optimize their applications by utilizing GPU architecture as well as the cloud capabilities of efficiency, scalability, and fast accessibility. For example, the Amazon EC2 cloud allows for taking advantage of the parallel performance of NVidia Telsa GPUs using the CUDA (NVidia) and OpenGL programming models for GPU computing.

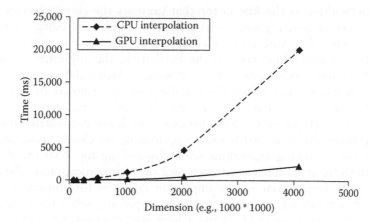

Figure 16.2 DEM interpolation with GPU and CPU.

16.3.4 **Remaining problems and future research**

Based on the discussions of DEM interpolation solutions, both CPU and GPU computing have the potential to address different computing intensive problems in geosciences. Cloud computing is able to provision flexible and powerful CPU-based and GPU-based computing resources. However, there are many remaining issues for cloud computing to support scientific computing. For example, Huang et al. (2013) found that cloud computing could provide a potential solution to support on-demand, small-sized, HPC applications. More research should be conducted to explore the feasibility of utilizing cloud computing to better support large-scale scientific computing and to learn how to best adapt to this new computing paradigm.

- *Interoperable and intelligent framework*—Both CPU and GPU computing can support parallel processing, and have their advantages and drawbacks in handling the geospatial computing tasks (Li et al. 2013). It poses grand challenges to make full use of these two computing paradigms to solve scientific computing problems. For example, it is a challenging issue to efficiently dispatch subtasks onto appropriate computing devices (e.g., CPU, GPU, or hybrid GPU and CPU) with different levels of computing capabilities. An interoperable and intelligent framework is urgently needed and should be designed and implemented to address this issue. Within this framework, a computing task is able to intelligently select computing devices and be completed efficiently.
- *Networking*—Huang et al. (2013) argued that cloud services in general are not fully ready to serve as a large-scale virtual cluster to support computing- and communication-intensive applications, and virtualized

networking is the key factor that prevents the cloud services from achieving better performance. Traditional HPC is being tuned for virtualized resources in the cloud computing era. For example, with the virtualized network as the bottleneck, the applications can try to perform redundant computing while reducing the communication overhead to fully utilize the scalable computing power of clouds. At the same time, cloud services can also address the network bottlenecks by (1) providing cloud instances with high-speed interconnection (Ostermann et al. 2010) and (2) optimizing the cloud service middleware scheduling algorithms while dispatching the VMs on physical hosts for HPC applications to reduce the communication overhead.

- *Cost*—Cost is one of the important concerns for consumers when adopting cloud computing solutions, especially when the public and hybrid clouds are used (Zhang, Cheng, and Boutaba 2010). Depending on the compute, storage, and communication requirements of an application, public cloud resources could be more or less expensive than hosting the application in-house for a long term (Armbrust et al. 2010). However, many applications have disruptive computing requirements where a large amount of computing resources should be leveraged when needed, such as real-time dust storm forecasting support (Huang et al. 2013). In this situation, it would be more cost-efficient to use a cloud service. A reasonable cost model should be proposed to capture dynamic resource requirements of an application and recommend to cloud consumers the best solutions to run applications (Chapter 6).

16.4 CONCURRENT INTENSITY

16.4.1 Introduction

Many geoscience applications face issues of concurrent access when many users are accessing applications simultaneously. Both hardware and software could be affected by redundant concurrent access. In some cases, the concurrent access will cause hardware issues such as out of memory or high CPU usage. In other cases, the concurrent access will cause software issues such as deadlock. The issue of concurrent access is critical for geoscience applications with large user communities such as GEOSS Clearinghouse (CLH) (Chapter 8)[1] and Geospatial Platforms[2] because: either (1) users may not get correct responses when a computer is implementing redundant requests; or (2) users wait too long to get the responses from the application.

---

[1] See GEOSS Clearinghouse at http://clearinghouse.cisc.gmu.edu/geonetwork.
[2] See Geospatial Platform at http://www.geoplatform.gov/home/.

16.4.2 An example with GEOSS Clearinghouse

CLH is a geoscience application, which has large numbers of globally distributed users. It provides functions of both local search and remote search. By analyzing the access logs, remote search with Open Geospatial Consortium (OGC) Catalog Service for the Web (CSW) is used by most users. Figure 16.3 shows the spatial distribution of the CLH users. There is a strong pattern where regions with high population densities (e.g., United States, Europe) have a large number of end users and generate lots of concurrent access. Figure 16.4 shows the user access frequency in Australia and New Zealand. The access frequency increases and decreases according to the change of time with spikes in specific hours when CLH has to handle a large number of concurrent end user requests. Therefore, concurrent intensity is critical to CLH.

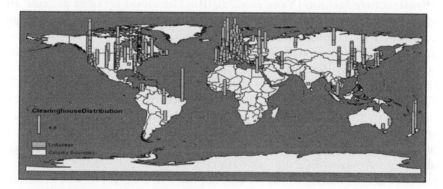

Figure 16.3 Users' distribution of CLH.

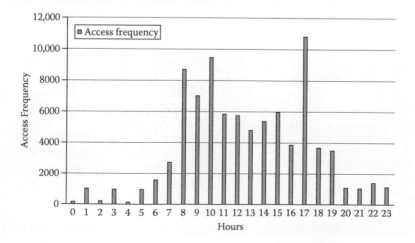

Figure 16.4 Australia and New Zealand users' access frequency of CLH.

16.4.3 Solution

There are several solutions that can be used to handle concurrent intensity in geoscience applications.

16.4.3.1 Global content delivery

Using cloud services, consumers may solve the issues of concurrent access in two ways—load balancing and elasticity. The load balancing and scalability of cloud computing can increase computing power with a low cost. CLH is replicated to balance the load from the massive number of end users. These replications are distributed at multiple locations and configured to function as a single service in a global geospatial cyber-infrastructure fashion (Yang et al. 2010). The cloud services provide an ideal platform to implement this load balancing mechanism. An image containing the configured application could be built in cloud services, and then a new replicated application can be launched in available cloud zones. Available cloud zones are listed in detail in Chapter 11. Currently, the CLH has instance images on Windows Azure and Amazon EC2. With the support of cloud services, we can easily build a load balancing mechanism to handle massive concurrent requests for CLH and other spatial services.

At present, most cloud providers have many regional data centers available (e.g., Amazon EC2 has 11 data centers as of early 2013). Based on the user access distribution, we can deploy five EC2 servers in different locations to handle global user access intensity. CLH could be deployed in multiple regions with EC2 cloud services (Figure 16.5). For example, when users access the services, the west coast users do not need to go to the east coast server.

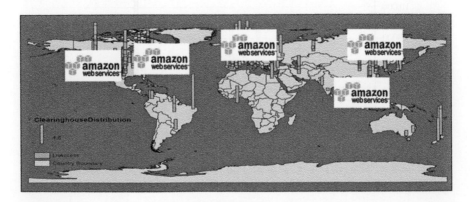

Figure 16.5 (See color insert.) Global content delivery to handle global user access.

16.4.3.2 Elasticity

Elasticity of cloud services can handle disruptive and massive loads from end users. Many cloud services provide elasticity mechanisms. For example, Amazon EC2 allows cloud consumers to define certain elasticity rules to automatically scale up the number of computing instances when the application is having massive concurrent access. More computing instances can balance the load and support more concurrent requests. In reality, cloud elasticity is widely applied in e-commerce, mobile, and Web applications.

Figure 16.6 uses CLH as an example to illustrate how the cloud elasticity responds to massive numbers of concurrent user requests. The figure spins off new EC2 cloud instances when a user request number exceeds a certain limit, with the x-axis value showing the user access number and the y-axis showing the system response time. The example illustrates a varying number of requests to the CLH. A new instance will be added to run CLH when the concurrent number increases and CPU usage is more than 90%. It is observed that when more computing instances are utilized, higher gains in performance can be obtained, and the response time can be kept to around 4 seconds with elastic computing resources (indicated by the five scaling instances). The elastic automated provision and release of computing resources greatly prepared us to respond to concurrent access spikes, meanwhile reducing cost by sharing computing resources with other applications when there are no concurrent access spikes.

16.4.3.3 Spatiotemporal indexing

A proper index mechanism will greatly improve the data retrieval speed. CLH built a specific indexing mechanism to reduce the execution time of a query thereby reducing the response time for user requests. With a

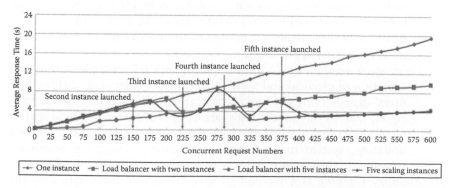

Figure 16.6 CLH response performance comparison by single-, two-, and five-load instances, and five auto-scaling instances. (From Yang et al., 2011a.)

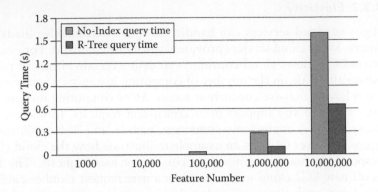

Figure 16.7 (See color insert.) Spatial indexing to improve the access speed.

shorter response time, CLH handles the increasing number of concurrent end users' requests. Figure 16.7 shows the query performance with and without an indexing mechanism. The indexing mechanism greatly speeds up the query performance of GEOSS Clearinghouse for handling more end user requests.

16.4.4 Remaining problems and future research

Data synchronization and performance—Cloud services can provide efficient capabilities such as load balancing and auto-scaling services to improve the concurrent performance for geoscience applications. However, many applications need to dynamically update data and contexts frequently, which causes a data inconsistency problem when configuring the load balancing and elasticity to launch instances at different times. In addition, there is a delay when starting a new instance using load balancing and auto-scaling services since it may take seconds or even minutes to boot up a VM. Hence, further investigations of these issues are necessary to improve the performance of these geospatial applications.

Using spatiotemporal principles—Spatiotemporal principles govern the evolution of physical and social phenomena (Yang et al. 2011a). End users' distributions and requests are affected by spatiotemporal principles, thereby having certain spatiotemporal patterns. These spatiotemporal patterns could help us better design the mechanism for data updating, system maintenance, and scaling computing resources, thereby increasing the capability to handle concurrent intensity.

Memory Caching is a method to save frequently used data into the computer RAM so that I/O performance can be improved. However, limited RAM space is the bottleneck for the cache mechanism.

The supercomputers with high computing power and big RAM size provide a potential solution to address this bottleneck. However, supercomputers are usually expensive, and it is helpful to develop a cost model for cloud consumers to balance the cost and system performance requirement. In addition, predicting and prefetching "mostly accessed data" could improve the cache usage rate, thereby improving the capability to handle concurrent intensity.

16.5 SPATIOTEMPORAL INTENSITY

The spatiotemporal intensity is fundamental for geosciences and contributes to other intensities (Yang et al. 2011a) as reflected in the following examples.

- *Spatiotemporal data*—Most geoscience data are recorded as a function of space–time dimensions either with static spatial information at a specific time stamp, or with changing time and spatial coverage (Terrenghi et al. 2010). For example, the daily temperature range for a specific place in the past 100 years is constrained by the location (place) and time (daily data for 100 years).
- *Data collection*—The advancement of sensing technologies increased our capability to measure more accurately and obtained better spatial coverage in a more timely fashion (Goodchild 2007). All datasets recorded for geosciences are spatiotemporal in either explicit (dynamic) or implicit fashion (static). The increase in resolution will greatly increase data volume and reveal much more geophysical principles, which is to be discovered by scientific research.
- *Natural phenomena and scientific research*—The study of geoscience phenomena has been described as space–time or geodynamics (Hornsby and Yuan 2008). In relevant geoscience studies such as atmospheric and oceanic sciences, the space–time and geodynamics have always been at the core of the research domains. And this core is becoming critical in almost all domains of pursuant human knowledge (Su et al. 2010).
- *Emergency response system*—A real-time response system could possibly involve all kinds of intensities: (1) massive spatiotemporal data that are historically recorded or collected in real time would be utilized, (2) the data will be processed and analyzed with high computing capacity to support better decision making, and (3) the potential for simultaneous massive public access. For example, millions of users accessed the weather.com Web site to keep updated with the progress of the hurricane during the 2012 Hurricane Sandy event.

Table 16.1 The Spatiotemporal Intensities Directly Contribute to Different Aspects
of the Intensities of Data, Computation, and Concurrent Access

	Data Intensity	Computation Intensity	Concurrent Intensity
Spatiotemporal Data	x	—	—
Data Collection	x	—	—
Phenomena Research	x	x	—
Emergency Response	x	x	x

Note: An "x" in the cell means the task has a specific intensity, for example, spatiotemporal data
has the characteristic of data intensity.

The four aspects for spatiotemporal intensity contribute directly to
the intensities of data, computation, and concurrent access (Table 16.1).
Therefore, the computing solutions for the other three intensities can also
be utilized to tackle the spatiotemporal intensive problems.

16.6 SUMMARY

This chapter laid out the grand challenges that geoscience faces in the 21st
century—the intensiveness of data, computing, concurrent access, and
spatiotemporal analysis. Yang et al. (2011b) argue that the latest advancements of cloud computing provide a potential solution to address these
grand challenges. Through examples of data intensity, computing intensity,
concurrent access intensity, and spatiotemporal intensity, we illustrate that
cloud computing is critical in their capabilities to: (a) enable the processing
of the distributed, heterogeneity, and big geoscience data (Section 16.2);
(b) support the computing intensive geoscience processing (Section 16.3);
(c) enable the timely response to worldwide distributed and locally clustered
users (Section 16.4); and (d) facilitate spatiotemporal intensities process and
applications (Section 16.5).

16.7 PROBLEMS

1. What are the computing challenges for data, computing, and concurrent and spatiotemporal intensity problems?
2. What are general solutions to address the data intensity issues?
3. How can cloud computing provide a potential solution to address
   data intensity problems? What are the remaining issues?
4. What is parallel computing? Explain the process of DEM interpolation using parallel computing.
5. What is graphics process unit (GPU) computing? What are the general steps for using GPU for DEM interpolation?

6. Summarize how cloud computing clouds provide a potential solution to address computing intensity problems. What are the remaining issues?
7. How is elasticity used to support concurrent intensity issues?
8. How can cloud computing provide a potential solution to address concurrent intensity problems? What are the remaining issues?
9. Enumerate four examples that include spatiotemporal intensity.

REFERENCES

Armbrust, M., A. Fox, R. Griffith, A. D. Joseph et al. 2010. A view of cloud computing. *Communications of the ACM* 53, no. 4: 50–58.

Armstrong, M. P., M. K. Cowles, and S. Wang. 2005. Using a computational grid for geographic information analysis: A reconnaissance. *The Professional Geographer* 57, no. 3: 365–375.

Audenino, P., L. Rognant, J. M. Chassery, and J. G. Planes. 2001. Fusion strategies for high resolution urban DEM. In *Remote Sensing and Data Fusion over Urban Areas, IEEE/ISPRS Joint Workshop* 2001, pp. 90–94. IEEE.

Buck, J. B., N. Watkins, J. LeFevre et al. 2011. SciHadoop: Array-based query processing in hadoop. In *Proceedings of 2011 International Conference for High Performance Computing, Networking, Storage and Analysis*, p. 66. ACM.

Cartwright, D. E. 2000. *Tides: A Scientific History*. United Kingdom: Cambridge University Press.

Donner, R., S. Barbosa, J. Kurths, and N. Marwan. 2009. Understanding the Earth as a complex system—Recent advances in data analysis and modelling in Earth sciences. *The European Physical Journal Special Topics* 174, no. 1: 1–9.

Drebin, R. A., L. Carpenter, and P. Hanrahan. 1988. Volume rendering. In *ACM Siggraph Computer Graphics* 22, no. 4, pp. 65–74. ACM.

Duffy, D. Q., J. L. Schnase, T. L., Clune, E. J. Kim, S. M. Freeman, J. H. Thompson, ... and M. E. Theriot. 2011. http://ntrs.nasa.gov/archive/nasa/casi.ntrs.nasa.gov/20120009187_2012009164.pdf (accessed August 9, 2013).

Dumbill, E. 2012. What Is Big Data? An Introduction to the Big Data Landscape. http://strata.oreilly.com/2012/01/what-is-big-data.html (accessed January 25, 2013).

Fox, G., S. H. Bae, J. Ekanayake, X. Qiu, and H. Yuan. 2010. Parallel data mining from multicore to cloudy grids. *High Speed and Large Scale Scientific Computing* 18: 311.

Huang, Q. and C. Yang. 2011. Optimizing grid computing configuration and scheduling for geospatial analysis: An example with interpolating DEM. *Computers & Geosciences* 37, no. 2: 165–176.

Huang, Q., C. Yang, K. Benedict et al. 2012. Using adaptively coupled models and high-performance computing for enabling the computability of dust storm forecasting. *International Journal of Geographic Information Science.* doi:10.1080/13658816.2012.715650.

Huang, Q., C. Yang, K. Benedict et al. 2013. Enabling dust storm forecasting using cloud computing. *International Journal of Digital Earth.* doi:10.1080/17538947.2012.749949.

Kumar, S. V., C. D. Peters-Lidard, Y. Tian et al. 2006. Land information system: An interoperable framework for high resolution land surface modeling. *Environmental Modelling & Software* 21, no. 10, 1402–1415.

Lämmel, R. 2008. Google's MapReduce programming model—Revisited. *Science of Computer Programming* 70, no. 1, 1–30.

Li, J., Y. Jiang, C. Yang, and Q. Huang. 2013. 59, 9: 78–89. Utilizing GPU and CPU for accelerating visualization pipeline. *Computers & Geosciences* 59, 9: 78–89.

Li, Z., C. Yang, M. Sun et al. 2013. A high performance Web-based system for analyzing and visualizing spatiotemporal data for climate studies. In *Web and Wireless Geographical Information Systems*, pp. 190–198. Heidelberg: Springer Berlin.

Li, Z., C. Yang, H. Wu, W. Li, and L. Miao. 2011. An optimized framework for seamlessly integrating OGC web services to support geospatial sciences. *International Journal of Geographical Information Science* 25, no. 4: 595–613.

Ostermann, S., A. Iosup, M. N. Yigitbasi et al. 2010. A Performance Analysis of EC2 Cloud Computing Services for Scientific Computing. Lecture Notes of the Institute for Computer Sciences: Social Informatics and Telecommunications Engineering 34, no. 4: 115–131. doi: 10.1007/978-3-642-12636-9_9.

Schuster, A. 1897. On lunar and solar periodicities of earthquakes. *Proceedings of the Royal Society of London* 61, no. 369–370: 455–465.

Singh, G., S. Bharathi, A. Chervenak et al. 2003. A metadata catalog service for data intensive applications. In *Supercomputing, 2003 ACM/IEEE Conference*, pp. 33–33. IEEE.

Theoderidis, Y., M. Vazirgiannis, and T. Sellis. 1996. Spatio-temporal indexing for large multimedia applications. In *Multimedia Computing and Systems. Proceedings of the 3rd IEEE International Conference*, pp. 441–448.

Wang, J. and Z. Liu. 2008. Parallel data mining optimal algorithm of virtual cluster. In *Fuzzy Systems and Knowledge Discovery. FSKD'08. 5th International Conference*, 5, pp. 358–362. IEEE.

Yang, C., M. Goodchild, Q. Huang et al. 2011a. Spatial cloud computing—How can geospatial sciences use and help to shape cloud computing. *International Journal of Digital Earth* 4, no. 4: 305–329.

Yang, C., R. Raskin, M. Goodchild, and M. Gahegan. 2010. Geospatial cyberinfrastructure: Past, present and future. *Computers, Environment and Urban Systems* 34, no. 4: 264–277.

Yang, C., M. Sun, K. Liu et al. 2013. Contemporary computing technologies for processing big spatiotemporal data. In *Space-Time Integration in Geography and GIScience: Research Frontiers in the U.S. and China*. Mei-Po Kwan, Douglas Richardson, Donggen Wang and Chenghu Zhou (eds), Dordrecht: Springer (in press).

Yang, C., H. Wu, Q. Huang, Z. Li, and J. Li. 2011b. Using spatial principles to optimize distributed computing for enabling the physical science discoveries. *Proceedings of the National Academy of Sciences* 108, no. 14: 5498–5503.

Yee, K. P., K. Swearingen, K. Li, and M. Hearst. 2003. Faceted metadata for image search and browsing. In *Proceedings of the SIGCHI Conference on Human Factors in Computing Systems*, pp. 401–408.

Zhang, Q., L. Cheng, and R. Boutaba. 2010. Cloud computing: State-of-the-art and research challenges. *Journal of Internet Services and Applications* 1, no. 1: 7–18.

Zhao, H., S. Ai, Z. Lv, and B. Li. 2010. Parallel accessing massive NetCDF data based on MapReduce. In *Web Information Systems and Mining*, pp. 425–431. Heidelberg: Springer Berlin.

Chapter 17

Cloud computing research for geosciences and applications

*Chaowei Yang, Qunying Huang, Zhipeng Gui,
Zhenlong Li, Chen Xu, Yunfeng Jiang, and Jing Li*

Cloud computing provides capabilities and solutions that transcend organizations, jurisdictional boundaries, and continents for many unsolved global challenges in the 21st century (NRC 2011b). However, the eventual success of cloud computing for geospatial sciences will be measured as if cloud computing can speed up geoscience research and improve geospatial application operations from a management perspective by, for example, cost savings (Luftman and Zadeh 2011). A lot of research and development is still needed to enable the solutions from at least the vision, technology, and social aspects (Yang, Xu, and Nebert 2013).

17.1 EVOLVING 21ST CENTURY VISION FOR GEOSCIENCE APPLICATIONS

Computer science advancements have been driven by various computing demands from noncomputer science domains. The advancement of cloud computing is no exception. Many 21st century geoscience challenges pose new requirements for cloud computing. Therefore, it is likely that geosciences will be a primary force driving the evolution of cloud computing. This section discusses the needs from three areas: fundamental geoscience research, interdisciplinary geoscience, and geoscience applications.

17.1.1 Fundamental geospatial science inquiries

Fundamental geoscience research is at the forefront of human knowledge pursuant and key to addressing many scientific and application challenges (NRC 2012b). For example, studying the Earth's geologic history may help scientists uncover solutions to help forecast many natural disasters, such as earthquakes and tsunamis (NRC 2011a, 2011b). Breakthroughs in fundamental research require scientists across the globe to share knowledge of their localized geospatial data gathered. Cloud computing has the potential to facilitate data processing, simulation, exchange, and collaboration. The global integration of various resources should be researched for cloud services to support global science discovery.

17.1.2 Integrating geoscience with other domains of science for new discoveries

McGrath (2011) argued that the most important scientific discoveries are happening now, but within, and especially across seemingly unrelated domains of science. The integration of science domains will require unprecedented tools and methodologies to facilitate integrated scientific discoveries. Cloud computing has the potential to facilitate scientific domain integration with a computing service that transcends each domain. Identifying the computing infrastructure needs for each domain and cross-domains will help in the planning of optimized cloud services to support the integration for new discoveries.

17.1.3 Application vision

Application vision is also a driver for technological advancements. For example, the Gore Digital Earth vision drove advancements in hardware, software, information management, and visualization, and eventually resulted in the highly successful Google Earth and Virtual Earth/Bing Maps. Continuing from the Digital Earth vision, smart Earth calls for real-time fusion of sensor observations and historical data to provide intelligent decision support for emergency responses (Liu, Xie, and Peng 2009; Liu et al. 2010). This massive real-time fusion of sensor information and historical data poses significant challenges for cloud computing.

To simulate a virtual environment for geospatial sciences, Virtual Geographic Environment (VGE) (Lin, Chen, and Lu 2011) was proposed to address those challenges. In turn, the project posed many challenges in visualization and user interaction for cloud computing to deal with big data and massive concurrent user access. Global military conflict simulators may provide another vision that drives one of the highest cost man-made events—the combat to solve disputes. The simulator could be used to train soldiers to help make smart decisions. The simulator may also provide a simulation mechanism to help reduce casualties and asset losses in a real combat environment (Tanase and Urzica 2009). Spatiotemporal information management poses significant challenges to the reliability and timeliness of cloud computing. Quickly developed emergency response maps, as described in Chapter 1, would help save hundreds of human lives and billions of dollars in asset loss. Therefore, it is important that a new mapping process is able to provide data, analyses, and mapping in a timely fashion. Significant challenges that exist within the new mapping process must be solved by new cartographic technologies (Meng 2011). Cloud computing has the potential to be the technological solution to many of the mapping challenges.

The continuing vision from geospatial domains will help set up the objective and drive technology advancements.

17.2 TECHNOLOGICAL ADVANCEMENTS

17.2.1 Cloud evaluation and selection

Evaluating and selecting cloud services is a challenge for cloud customers. To resolve this issue, the following research aspects need to be addressed:

- Cloud service measurement criteria are the foundation for measuring, evaluating, and selecting cloud services. Although various measurement criteria have been proposed (Kulkarni 2012; Repschläger et al. 2011), these criteria are not well defined. For example, although the National Institute for Standards and Technology (NIST) explored open issues and recommendations (Badger et al. 2012), established the RATAX Cloud Metrics Sub Group,[1] and standardized the criteria and associated models, further efforts are still needed to define consistent, reusable, and operational models to support comprehensive measurements of cloud services (NIST 2012).

- Return On Investment (ROI) is a significant determining factor in cloud selection from the cloud consumers' perspective. Cost calculation models[2] (Li et al. 2009) have been developed for calculating cloud Total Cost of Ownership (TCO) and Utilization Cost. To help customers understand their ROI and make wise selections, deliberate cost models should be developed to make accurate an estimation and analysis on the potential cost composition and utilization of imbalance factors.

- The third-party auditing is essential to provide a trustable understanding of the performance, reliability, and consistency of cloud services. CloudSleuth developed a monitoring platform called Global Provider View[3] to collect and visually analyze the performance and availability of the many popular cloud services. It continuously monitors the top cloud services around the world by deploying and running a sample Web application on each of the cloud services. For broad monitoring for further evaluation, the architecture of monitoring services

---

[1] See NIST at http://collaborate.nist.gov/twiki-cloud-computing/bin/view/CloudComputing/RATax_CloudMetrics.

[2] See Microsoft Windows Azure Platform TCO Calculator at http://www.microsoft.com/brasil/windowsazure/tco/.

[3] See Global Provider View at https://cloudsleuth.net/global-provider-view.

needs to be well designed and various sample applications should be considered for different scenarios.

- To assist cloud customers to understand the advantages and disadvantages of cloud services as well as to make a wise selection, advanced selection principles and models should be investigated. Based on these selection methods, sophisticated and user-friendly cloud advisory tools (Goscinski et al. 2010; Martens et al. 2011) and Web sites[1] should be developed. Besides collecting, comparing, and visually analyzing cloud service features (e.g., capacities, cost fee, and user feedback), these tools and Web sites should be capable of making recommendations based on consumer requirements and preferences. For example, Andrzejak (2010) proposed a probabilistic model for optimizing monetary costs, performance, and reliability, given user and application requirements and dynamic conditions. This model can help consumers bid optimally on Amazon Spot Instances to reach desired objectives. Furthermore, to optimize the cloud service selection based on the Service Level Agreement (SLA), cost, application features and spatiotemporal characteristics of application users/cloud services/data and other properties, optimization algorithms should also be developed.

17.2.2 Cloud service resource management

Cloud resources (e.g., data and computing centers) represent a significant investment in capital outlay and ongoing costs. Optimized cloud resource management and utilization cannot only improve resource utilization and performance, but also reduce the budget, energy, and labor, for both cloud providers and cloud consumers.

Resource stranding and fragmentation seriously obstruct the utilization of computing resources and also increase the management cost. Increasing network agility and providing appropriate incentives can shape resource consumption and improve computing resource utilization (Greenberg et al. 2009). To reduce the cost raised by networks of geographically dispersed data centers, the joint optimization of network and data center resources and new mechanisms for the geodistributing state should also be proposed.

Saving electricity used on computing resources is important to reduce global energy consumption. Beloglazovet et al. (2010) proposed an energy efficient resource management system for virtualized cloud data centers to reduce operational costs and provide required Quality of Service (QoS). Energy savings can be achieved by continuous consolidation of Virtual Machines (VMs) according to the current utilization of resources, the virtual network topologies established between VMs, and the thermal state of computing nodes.

---

[1] See Cloud Computing Providers at http://cloud-computing.findthebest.com/.

Automating management of VMs can reduce labor cost and time for cloud providers. To control the virtualized environment, an intelligent autonomic resource manager can decouple provisioning resources from the dynamic placement of VMs. By using a constraint programming approach, the manager can optimize the global utility by considering both the degree of SLA fulfillment and the operating costs.

17.2.3 Data backup and synchronization

Cloud storage services are changing the way people access and store data. There are a variety of cloud storage services, such as Amazon S3, Cloud Drive, Dropbox, Box, SugarSync, SkyDrive, CloudApp, Google Driver, and Apple iCloud as introduced in Chapter 3, Section 3.3. Cloud storage services provide a powerful capability for backup and synchronization among different terminals for our daily data and information, such as videos, music, e-mails, and documents. However, backup and synchronization of data becomes the scalability bottleneck for scientific applications to best utilize on-demand computing power. In the cloud, multiple servers can be easily deployed for Web applications to balance the access requests. However, there are difficulties with concurrent data access. Consequently, multiple servers have to access the data store sequentially (Das, Agrawal, and Abbadi 2009).

Data backup and synchronization is one of the most important research issues within cloud computing. Das, Agrawal, and Abbadi (2009) proposed ElasTraS with the target to address the issue of scalability and elasticity of database transactions in the cloud. Vrable, Savage, and Voelker (2009) presented a system named *Cumulus* for efficiently implementing file system backups over the Internet, specifically designed under a thin cloud assumption. Toka, Dell'Amico, and Michiardi (2010) studied the benefits of a peer-assisted approach for online backup applications. However, the methodologies for data backup and synchronization with massive geospatial data and frequent operations need to be further explored, and the cloud providers should integrate those strategies into their cloud services that can be easily configured by the cloud consumers.

17.2.4 Interoperability

Interoperability has to rely on the standards developed by different organizations, such as OGC, OGF, NIST, ISO, and IEEE, through a systematic architecture design to support sharing of distributed computing resources at different levels (NIST 2012). This has to be driven by large user groups, application domains, vendors, and governments to achieve the required level of interoperability (Lee 2010).

With many cloud service models (e.g., IaaS, PaaS, and SaaS) and providers (e.g., Amazon and Windows), interoperability among different cloud services is of great interest for consumers to take full advantage of innovative technology. However, the lack of standards for cloud computing interoperability has become a critical impediment that hinders cloud computing evolution. Currently, cloud providers are using different implementation methods and interfaces. For example, Amazon uses Xen as the hypervisor for its EC2 platform, while the virtualization of Windows Azure is based on Hyper-V,[1] which was formerly known as *Windows Server Virtualization*. These differences lead to many interoperability issues such as vendor lock-in and application portability.

Fortunately, several organizations are starting to address these issues from different aspects and directions. NIST is leading the effort closely with standards communities, the private sectors, and other stakeholders including federal agencies to work on the roadmap and reference architecture of cloud computing standards (Hogan et al. 2011). A continuously updated "Inventory of Standards Relevant to Cloud Computing" is compiled by the NIST cloud computing standards working group, including volunteer participants from industry, government, and academia. This inventory of standards collects the highest-level protocols, definitions, and standards that are applicable widely to the cloud computing use cases.[2] The Distributed Management Task Force (DMTF)[3] has formed the open cloud standards incubator focusing on standardizing interactions between cloud services by developing cloud resource management protocols. Other organizations include Open Group Cloud Computing Work Group,[4] Open Grid Forum,[5] Open Cloud Consortium,[6] and Cloud Security Alliance.[7] While some progress has been made on the cloud interoperability issues, more collaborative efforts from the government, academia, and large cloud vendors are required.

Driven by the urgent need of an interoperable cloud computing environment, and led by NIST and other standard organizations, future research addressing the cloud interoperability problem includes but is not limited to:

- Developing and maturing standards to support interoperability for all cloud models from low-layer stack IaaS to high-layer stack SaaS. Specific standards are required for particular cloud service models.
- Developing standard mediator APIs to enable cloud consumers to utilize, manage, compare, and integrate cloud services from different

---

[1] See Hyper-V at http://en.wikipedia.org/wiki/Hyper-V.
[2] See NIST at http://collaborate.nist.gov/twiki-cloud-computing/bin/view/CloudComputing/StandardsInventory.
[3] See DMTF at http://www.dmtf.org/standards.
[4] See The Open Group at http://www.opengroup.org/getinvolved/workgroups/cloudcomputing.
[5] See Open Grid Forum at http://www.ogf.org/.
[6] See Open Cloud Consortium at www.opencloudconsortium.org.
[7] See Cloud Security Alliance at https://cloudsecurityalliance.org/.

cloud providers. For example, applications built on various cloud services can be used interoperably by using the standard mediator API.

- Developing cloud resource management protocols and security mechanisms to facilitate cloud interoperability. For example, orchestration layers can be used to build business processes and workflows using the cloud services provided by different cloud providers (Parameswaran and Chaddha 2009).

17.2.5 New visualization and interactive systems

Visualization and interactive systems have made great contributions in supporting scientific explorations of large-scale multidimensional data. Today's cloud-based computing infrastructure and data warehouse have greatly enhanced our capabilities of performing computing intensive modeling and simulations, as well as managing voluminous distributed data. In addition, they have offered new opportunities and challenges to develop visualization and interactive systems. Although various methods, techniques, and tools have been designed and developed to facilitate distributed visualization on massive data, these approaches may not be applicable in the cloud computing environment because of the complexities of cloud framework and supporting infrastructure. Future research should focus on:

- Customizing existing visualization algorithms and approaches with cloud services to support ultra-scale interactive visualization where computing intensity is always a bottleneck. For example, Amazon EC2 provides cluster instances with Graphics Processing Units (GPUs), which can be tailored to visualization and interactive systems.
- Exploring spatiotemporal principles in governing the implementations of visualization and interactive systems with cloud services. Comprehensive evaluations should be conducted to provide guidance for system level design.
- Designing an adaptive workflow that can best utilize cloud resources, from data preprocessing to final display, in a cloud environment. For example, how to allocate interactive rendering tasks based on cloud computing resources needs further study.
- Building a remote cloud visualization service to allow users from anywhere to interactively access the data, images, videos, and applications as services. The service should be able to support on-the-fly visualization of partial or intermediate results from data analyses. The objective is that cloud consumers do not have to worry about problems and limitations caused by different operation systems, physical machine locations, and their capabilities (Tanahashi et al. 2010).

17.2.6 Reliability and availability

Reliability and availability refer to being able to efficiently access the environment from different regions. Both are challenging to achieve in a cloud computing environment where most components are distributed and independently managed. Even large cloud providers have not overcome these challenges. In 2012, the two earliest cloud providers, Amazon and Google, had multiple outages of their cloud services, which led to service interruptions from network problems. For example, problems with Amazon's AWS in Northern Virginia knocked several popular services offline, including Netflix, on Christmas Eve.[1]

There are two possible reasons for the presence of service unavailability and unreliability, including (1) the servers, which consist of multiple hard disks, memory modules, network cards, and processors, may fail even though they are being carefully engineered (Venkatesh et al. 2010), and (2) services are delivered through the Internet, and therefore are vulnerable to network outages. The lack of reliability may also introduce big financial losses to the business for some users, especially large corporate users.

Many research efforts are increasingly put on reducing hardware and network failures to improve reliability of cloud computing. For example, Venkatesh et al. (2010) presented a model to proactively predict hardware failures for a physical machine, which can result in moving workload and data off such a server in time to avoid any possible service disruption. In order to obtain the confidence of large enterprises to shift computing styles to cloud computing, great effort should be put into investigating strategies for cloud system failure predication, response, and recovery from the perspectives of both the cloud providers and cloud consumers.

17.2.7 Real-time simulation and access

Real-time simulation and access is essential for different types of decision supports from emergency response to individual daily life. In addition, new devices, such as tablets and smartphones, are increasingly used by the public. This makes a real-time response system more easily accessible than ever before. On the other hand, it also introduces a tremendous workload for cloud services, especially during the emergent time period, where massive requests are possible (Huang et al. 2013). Cloud computing provides great opportunities to support such real-time systems with the requirement of a fast response. For example, Charalampos et al. (2010) implemented a healthcare information management system using cloud computing that enables electronic healthcare data storage, update, and

---

[1] See Forbes at http://www.forbes.com/sites/kellyclay/2012/12/24/amazon-aws-takes-down-netflix-on-christmas-eve/.

retrieval. More research should be conducted including theory-based simulation and multiscale, multicomponent modeling, as well as data intensive and interactive visualization capability for both cloud computing platforms and applications (NRC 2011a,b).

17.3 SYNERGISTIC ADVANCEMENT OF SOCIAL SCIENCE AND CLOUD COMPUTING

17.3.1 Cloud management

There are five cloud computing characteristics defined by NIST (2011) to depict features of cloud computing from different perspectives (Yang et al. 2011). A common requirement for implementing these characteristics is to automate the management process across the cloud computing infrastructure. At the operational level, user interventions should be minimized to ensure autonomous self-regulation of cloud computing. Self-regulation demands the breaking down of existing organizational restrictions to support the automatic processing (Choi and Lee 2010). At the technological level, standards should be created and supported by different services. More difficulties come from human factors; for example, could a cloud service be trusted by cloud consumers to host their data and applications?

17.3.2 Cloud outreach

Cloud computing emerges as a potential computing solution for local, regional, and global challenges. The solution will be a progressive process and needs deliberate communication with the public. For example, cloud computing was first advertised as a trailblazer for solving all computing problems, but it took Amazon four years (2006 to 2010) to mature its high performance computing offerings. The notion of utility computing makes the public, or even some information technology decision makers, believe that we only need 100 to 200 computing centers across the United States and all the computing needs can be served from the centers. But the fact is that cloud computing will be limited by network bandwidth, and the spatiotemporal principles constrain us to collocating our data with computing infrastructure, therefore, proximity has to be considered between cloud computing and data, users, problems, and applications. Another underestimated impact is security. Although a cloud system may be more secure from attacks, the overall security of data, privacy, and meeting government needs sometimes prevent governments from adopting the public cloud. Therefore, it is essential to convey the right message to the public about cloud computing so that it is neither overcommitted nor undercommunicated.

17.3.3 Security and regulations

Security and regulations are concerned with utilizing cloud computing for certain information and applications; for example, labor and social security applications (Lu 2010). Currently, security and regulations are among the biggest concerns when individuals or enterprises are considering adopting cloud services. Under the traditional IT environment, consumers have complete control of their IT systems, including (1) the location of the data center, (2) permission for who can access and use these infrastructure, data, and services, and (3) security strategies. With cloud services, enterprises are losing the physical control of the IT systems. For example, an enterprise may be prohibited from hosting their data and information in a foreign country. Therefore, companies must pay attention to the available regions provided by cloud providers when selecting a cloud service, and check whether there is a conflict between the uses of these cloud services and security regulations. For example, all federal IT applications must meet certain security requirements of the Federal Risk and Authorization Management Program (FedRAMP).[1] FISMA requires federal agencies to develop, document, and implement an information security system for its data and infrastructure. In addition to the national regulations, some companies also need to follow the industry rules and standards. For example, all organizations, large or small, in the United States, need to comply with the Sarbanes-Oxley Act (SOX).[2] Since a client can log in from any location to access data and applications, it is possible the client's privacy could be compromised within a cloud computing environment. Several mechanisms can be applied including, for example, using authentication techniques, such as user names and passwords and each user can access only the data and applications relevant to his or her job.

Improvements to cloud service security and privacy levels are needed for data and services including a secure Web environment for developing applications (O'Leary and Kaufman 2011), creation of international license agreements or exceptions to ensure that export-controlled technical data stored on the cloud is secure and protected (Schoorl 2011), and protection of sensitive data, privacy, and systems while maintaining the sharing spirit of cloud computing. Under such a circumstance, it is very important that cloud providers enable consumers or third parties to review the security and privacy policy and verify its completeness and validity. More research, such as data encryption and decryption solutions, should be conducted to secure confidentiality of data and information (Patial and Behal 2012).

---

[1] See GSA at http://www.gsa.gov/portal/category/102371.
[2] See the Sarbanes-Oxley Act at http://www.soxlaw.com.

17.3.4 Global collaboration

Globalization enabled partially by digital technology has changed the landscape of research collaboration. Collaboration at the global scale has been vital for the solutions of many challenges, such as climate change, poverty, and sustainability. While these challenges have their manifestations locally, they have global impacts. For example, companies seeking to maximize their profits may move their manufacturing lines to places with lower labor costs, and create development as well as pollution there. Some pollution can affect global regions. Global collaboration is hence indispensable for the effectiveness of a solution, and might be the only way for a solution. Cloud computing started as a promising paradigm to a collaborative web of research to support a global collaboration on several fronts: (1) The integration of multiple scientific domains through common information infrastructure (Yang et al. 2010). An integrated environment simplifies data sharing, which is crucial for scientific research at the global level. (2) Cloud computing can bring new opportunities from the identified direction of Internet-of-Things (IoT) for knowing where everything is at all times to advanced GIScience. IoT has been a new way for data collection, and has generated new data types for research in environmental monitoring and urban planning. Cloud computing creates a new mechanism for connecting citizens through ubiquitous computing devices. The new mechanism has demonstrated its effectiveness in areas such as crowd fund raising and public participation in disaster response. Ubiquitous IoT and collaborative webs of citizens can help detect and track natural and social events. In the big data era, visual analytics has become an effective method for pattern detection. The process is data intensive as well as computing intensive. Cloud computing has already shown a capacity to meet those requirements (Yang et al. 2011b). Eventually we need to educate our next generation for addressing challenges in sustained development. Cloud computing suggests a new direction for education, where resources can be shared globally, knowledge can be distributed globally, and human intelligence can be accumulated for solving both local and global problems (Goodchild 2010). Craglia et al. (2012) also emphasized the importance of global collaboration in the light of many developments in information technology, data infrastructures, and Earth observation. It is essential to develop a series of collaborations at the global level to turn the vision into reality.

17.4 SUMMARY

This chapter reviews the potential of cloud computing in the context of solving 21st century challenges and the research needs for advancing cloud computing in the next decade. Section 17.1 introduces the evolution of missions needed for geoscience applications. Section 17.2 introduces technological advancements needed for cloud computing to support geoscience. And, Section 17.3 introduces the social science advancements needed.

17.5 PROBLEMS

1. Give an example of geoscience challenges and relevant discovery needs.
2. Give an example to illustrate how geoscience interacts with other domains of science.
3. What are the visions of geosciences and applications? What role do the visions play in cloud computing?
4. Enumerate four cloud computing aspects that need advancement and discuss the potential directions to achieve the advancements.
5. How could cloud computing help advance social sciences?
6. How can social science advancements help the cloud computing evolution?

REFERENCES

Andrzejak, A., D. Kondo, and D. P. Anderson. 2010. Exploiting non-dedicated resources for cloud computing. In *Network Operations and Management Symposium (NOMS)*, pp. 341–348. IEEE.

Badger, L., T. Grance, R. Patt-Corner, and J. Voas. 2012. Cloud computing synopsis and recommendations. *NIST Special Publication* 800, 146.

Beloglazov A. and R. Buyya. 2010. Adaptive Threshold-Based Approach for Energy-Efficient Consolidation of Virtual Machines in Cloud Data Centers. *Proceedings of the 8th International Workshop on Middleware for Grids, Clouds and e-Science (MGC 2010)*, November 29–December 3. Bangalore, India. ACM.

Choi, S. and G. Lee. 2010. 3D viewer platform of cloud clustering management system: Google Map 3D. *Communication and Networking*: 218–222.

Craglia, M., K. de Bie, D. Jackson et al. 2012. Digital Earth 2020: Towards the vision for the next decade. *International Journal of Digital Earth* 5, no. 1: 4–21.

Das, S., D. Agrawal, and A. E. Abbadi. 2009. Elastras: An elastic transactional data store in the cloud. *USENIX HotCloud* 2.

Goodchild, M. F. 2012. Twenty years of progress: GIScience in 2010. *Journal of Spatial Information Science* 1: 3–20.

Goscinski, A. and M. Brock. 2010. Toward dynamic and attribute based publication, discovery and selection for cloud computing. *Future Generation Computer Systems* 26, no. 7: 947–970.

Greenberg, A., J. Hamilton, D. A. Maltz, and P. Patel. 2008. The cost of a cloud: Research problems in data center networks. *ACM SIGCOMM Computer Communication Review* 39, no. 1: 68–73.

Hogan, M., F. Liu, A. Sokol, and J. Tong J. 2011. NIST cloud computing standards roadmap. *NIST Special Publication* 35.

Huang, Q., C. Yang, K. Benedict, S. Chen, A. Rezgui, and J. Xie. 2013. Utilize cloud computing to support dust storm forecasting. *International Journal of Digital Earth.* doi:/10.1080/17538947.2012.749949.

Huang, Q., C. Yang, D. Nebert, K. Liu, and H. Wu. 2010. Cloud computing for geosciences: Deployment of GEOSS Clearinghouse on Amazon's EC2. In *Proceedings of the ACM SIGSPATIAL International Workshop on High Performance and Distributed Geographic Information Systems*, pp. 35–38. ACM.

IEEE. 2011. IEEE technology time machine symposium on technologies beyond 2020, *TTM* 2011, p. 46.

Kim, I. H., M. Tsou et al. 2013. Enabling digital earth simulation models with cloud computing or grid computing? Two approaches to support web GIS simulation frameworks. *International Journal of Digital Earth* 6, no. 4: 383–403.

Kulkarni, G. 2012. Cloud computing—Software as a service. *International Journal of Cloud Computing and Services Science (IJ-CLOSER)* 1, no. 1: 11–16.

Lee, C. A. 2010. A perspective on scientific cloud computing. In *Proceedings of the 19th ACM International Symposium on High Performance Distributed Computing*, pp. 451–459. ACM.

Li, B., J. Li, J. Huai, T. Wo, Q. Li, and L. Zhong. 2009 (September). EnaCloud: An energy-saving application live placement approach for cloud computing environments. In *Cloud Computing. CLOUD'09, IEEE International Conference*, pp. 17–24. IEEE.

Lin, H., M. Chen, and G. Lu. 2011. Virtual geographic environment: A workspace for computer-aided geographic experiments. *Annals of AAG.* doi:10.1080/00 045608.2012.689234.

Lin, L., L. De-Ren, Y. Bo-Xiong, and L. Wan-Wu. 2010. Smart planet based on geomatics. In *Multimedia Technology (ICMT), International Conference*, pp. 1–4.

Liu, F., J. Tong, J. Mao, R. Bohn, J. Messina, L. Badger and D. Leaf. 2010. NIST cloud computing reference architecture. *NIST Special Publication*, 500, 292.

Liu, C., Z. Xie, and P. Peng. 2009. A discussion on the framework of a smarter campus. In *Intelligent Information Technology Application. IITA. 3rd International Symposium*, 2, pp. 479–482.

Lu, X. 2010. Service and cloud computing oriented Web GIS for labor and social security applications. In *Information Science and Engineering (ICISE), 2010 2nd International Conference*, pp. 4014–4017.

Luftman, J. and H. S. Zadeh. 2011. Key information technology and management issues 2010–11: An international study. *Journal of Information Technology* 26, no. 3: 193–204.

Martens, B. and F. Teuteberg. 2011. Risk and Compliance Management for Cloud Computing Services: Designing a Reference Model. In *AMCIS*.

McGrath, E. 2011. *The Most Important Scientific Discovery of All Time: McGrath's Hypothesis: And the Scientific Revolution Going on Right Now*. Strategic Book Publishing, ISBN-13: 978-1612044897, p. 146.

Meng, L. 2011. Cartography for everyone and everyone for cartography—Why and how? *Kartographische Nachrichten* 61, no. 5: 246–253.

NIST. 2011. The NIST Definition of Cloud Computing. http://csrc.nist.gov/ publications/nistpubs/800-145/SP800-145.pdf (accessed December 29, 2012).

NIST. 2012. NIST Cloud Computing Reference Architecture Cloud Service Metrics Description (Draft). http://collaborate.nist.gov/twiki-cloud-computing/pub/CloudComputing/RATax_CloudMetrics/RATAX-CloudServiceMetrics Description-DRAFT-v1.1.pdf (accessed March 12, 2013).

NRC. 2011. *Grand Challenges in Earthquake Engineering Research: A Community Workshop Report.* Washington DC: National Academies Press, p. 23.

NRC. 2011b. *Assessment of Impediments to Interagency Collaboration on Space and Earth Science Missions.* Washington DC: National Academies Press, p. 67.

NRC. 2012. *International Science in the National Interest at the U.S. Geological Survey.* Washington DC: National Academies Press, pp. 161.

O'Leary, M. A. and S. Kaufman. 2011. MorphoBank: Phylophenomics in the "cloud." *Cladistics* 27, no. 5: 529–537.

Papamanthou, C., R. Tamassia, and N. Triandopoulos. 2010. Optimal authenticated data structures with multilinear forms. In *Pairing-Based Cryptography-Pairing.* Berlin, Heidelberg: Springer, pp. 246–264.

Parameswaran, A. V. and A. Chaddha. 2009. Cloud interoperability and standardization. *SETlabs Briefings* 7, no. 7: 19–26.

Patial, A. and S. Behal. 2012. RSA algorithm achievement with Federal Information Processing Signature for Data Protection in cloud computing. *International Journal of Computers & Technology* 3, no. 1: 34–38.

Repschläger, J., S. Wind, R. Zarnekow, and K. Turowski. 2011 (September). Developing a Cloud Provider Selection Model. In *EMISA*, pp. 163–176.

Schoorl, J. A. 2011. Clicking the export button: Cloud data storage and U.S. dual-use export controls. *Geo. Wash. L. Rev.* 80: 632.

Tanase, C. and A. Urzica. 2009. Global military conflict simulator. *Studies in Computational Intelligence* 237: 313–318.

Tanahashi, Y., C. K. Chen, S. Marchesin, and K. L. Ma. 2010. An interface design for future cloud-based visualization services. In *Cloud Computing Technology and Science (CloudCom), 2010 IEEE 2nd International Conference*, pp. 609–613. IEEE.

Toka, L., M. Dell'Amico, and P. Michiardi. 2010. Online data backup: A peer-assisted approach. *In Peer-to-Peer Computing (P2P), IEEE 10th International Conference*, pp. 1–10.

Van, N. H., D. F. Tran, and J. M. Menaud. 2009. Autonomic virtual resource management for service hosting platforms. In *Proceedings of the 2009 ICSE Workshop on Software Engineering Challenges of Cloud Computing, IEEE Computer Society*, pp. 1–8.

Vrable, M., S. Savage, and G. M. Voelker. 2009. Cumulus: File system backup to the cloud. *ACM Transactions on Storage (TOS)* 5, no. 4: 14.

Wen, Y., M. Chen, G. Lu, H. Lin, L. He, and S. Yue. 2013. Prototyping an open environment for sharing geographical analysis models on cloud computing platform 6, no. 4: 356–382.

Yang, C., M. Goodchild, Q. Huang et al. 2011a. Spatial cloud computing: How can the geospatial sciences use and help shape cloud computing? *International Journal of Digital Earth* 4, no. 4: 305–329.

Yang, C., Y. Xu, and D. Nebert. 2013. Redefining the Possibility of Digital Earth and Geosciences with Spatial Cloud Computing, *International Journal of Digital Earth* 6, no. 4: 297–312.

Yang, C., D. Nebert, and D. R. Fraser Taylor. 2011b. Establishing a sustainable and cross-boundary geospatial cyberinfrastructure to enable polar research. *Computers & Geosciences* 37, no. 11: 1721–1726.

Yang, C., R. Raskin, M. Goodchild, and M. Gahegan. 2010. Geospatial cyberinfrastructure: Past, present and future. *Computers, Environment and Urban Systems* 34, no. 4: 264–277.

Yue, P., H. Zhou, J. Gong, and L. Hu. 2013. Geoprocessing in cloud computing platforms—A comparative analysis. *International Journal of Digital Earth* 6, no. 4, 404–425.

Index